The Student's
Anatomy
of Stretching Manual

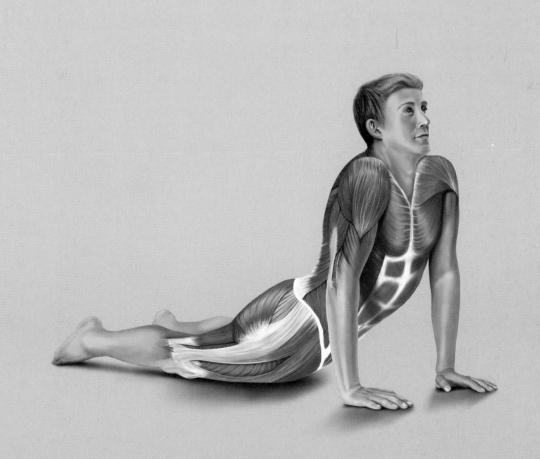

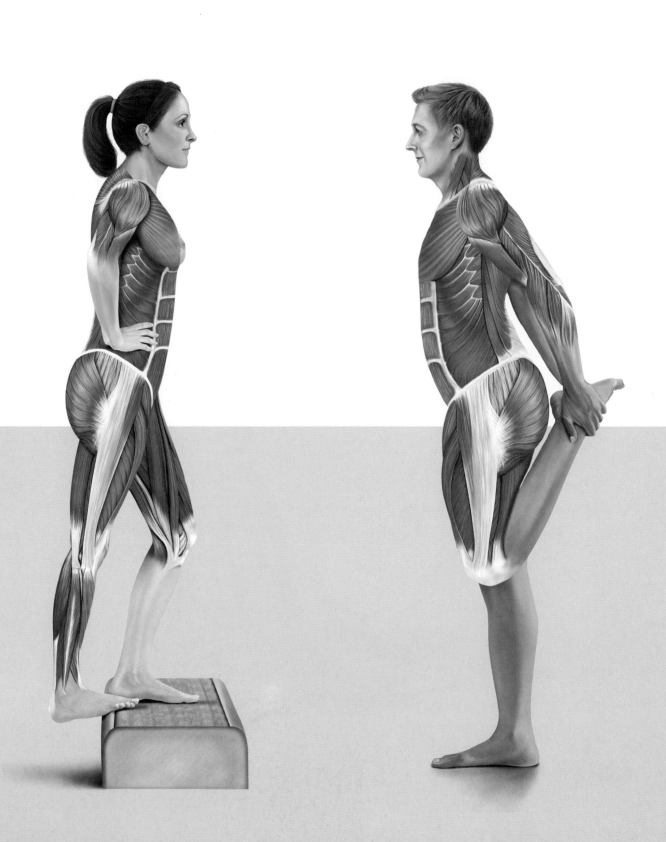

The Student's
Anatomy
of Stretching Manual

Chief consultant
Professor Ken Ashwell B.Med.Sc., M.B., B.S., Ph.D.

BARRON'S

First edition for the United States, its territories and dependencies, and Canada published in 2014 by Barron's Educational Series, Inc.

First published in 2014 by Global Book Publishing Pty Ltd
201 (Suite 9) Lakeside Corporate Centre
29-31 Solent Circuit, Baulkham Hills
NSW 2153, Australia
www.globalbookpublishing.com.au

All inquiries should be addressed to:
Barron's Educational Series, Inc.
250 Wireless Boulevard
Hauppauge, NY 11788
www.barronseduc.com

ISBN 978-1-4380-0391-7
Library of Congress Control Number: 2013939657
Printed in China by I-Book Printing Limited

9 8 7 6 5 4 3 2 1

It is recommended that anyone who is considering participating in an exercise program should consult a physician before starting and that no one should attempt a new exercise without the supervision of a certified professional. While every care has been taken in presenting this material, the anatomical and medical information is not intended to replace professional medical advice; it should not be used as a guide for self-treatment or self-diagnosis. Neither the authors nor the publisher may be held responsible for any type of damage or harm caused by the use or misuse of information in this book.

Publisher	James Mills-Hicks
Project Manager	Selena Quintrell
Editor	Lachlan McLaine
Chief Consultant	Ken Ashwell B.Med.Sc., M.B., B.S., Ph.D.
Authors	Ken Ashwell B.Med.Sc., M.B., B.S., Ph.D Tim Foulcher B.App.Sc., M.Phty. Michael Baker B.App.Sc., M.App.Sc., Ph.D., A.E.P.
Cover Design	Kylie Mulquin
Designer	Kylie Mulquin
Illustrator (Exercises)	Joanna Culley, B.A.(Hons), R.M.I.P., M.M.A.A., medical-artist.com
Illustrations Editor	Selena Quintrell
Editorial Coordinator	Chrysoula Aiello
Other Illustrations	David Carroll, Peter Child, Deborah Clarke, Geoff Cook, Marcus Cremonese, Beth Croce, Hans De Haas, Wendy de Paauw, Levant Efe, Mike Golding, Mike Gorman, Jeff Lang, Alex Lavroff, Ulrich Lehmann, Ruth Lindsay, Richard McKenna, Kristen W. Marzejon, Annabel Milne, Tony Pyrzakowski, Oliver Rennert, Caroline Rodrigues, Otto Schmidinger, Bob Seal, Vicky Short, Graeme Tavendale, Thomson Digital, Jonathan Tidball, Paul Tresnan, Valentin Varetsa, Glen Vause, Spike Wademan, Trevor Weekes, Paul Williams, David Wood

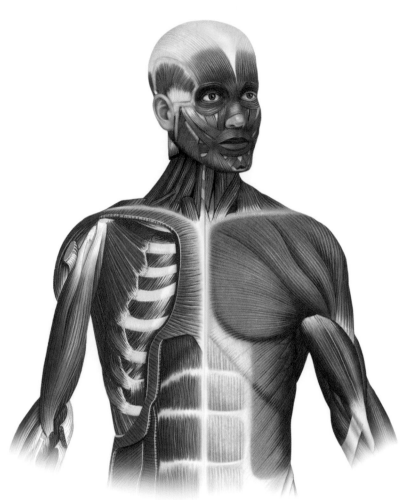

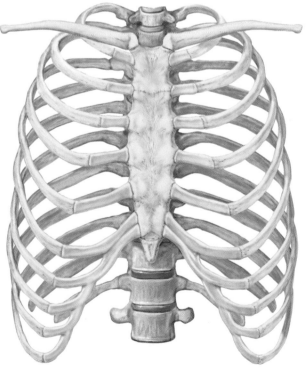

6

Contents

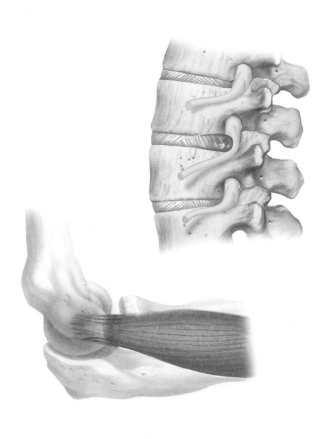

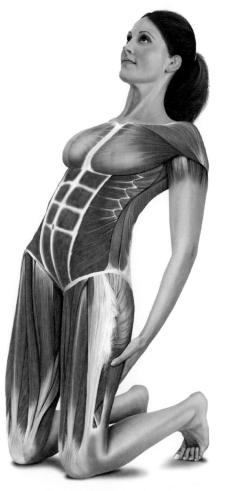

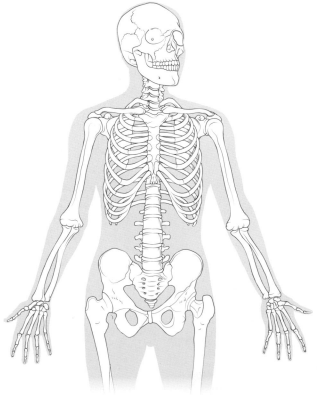

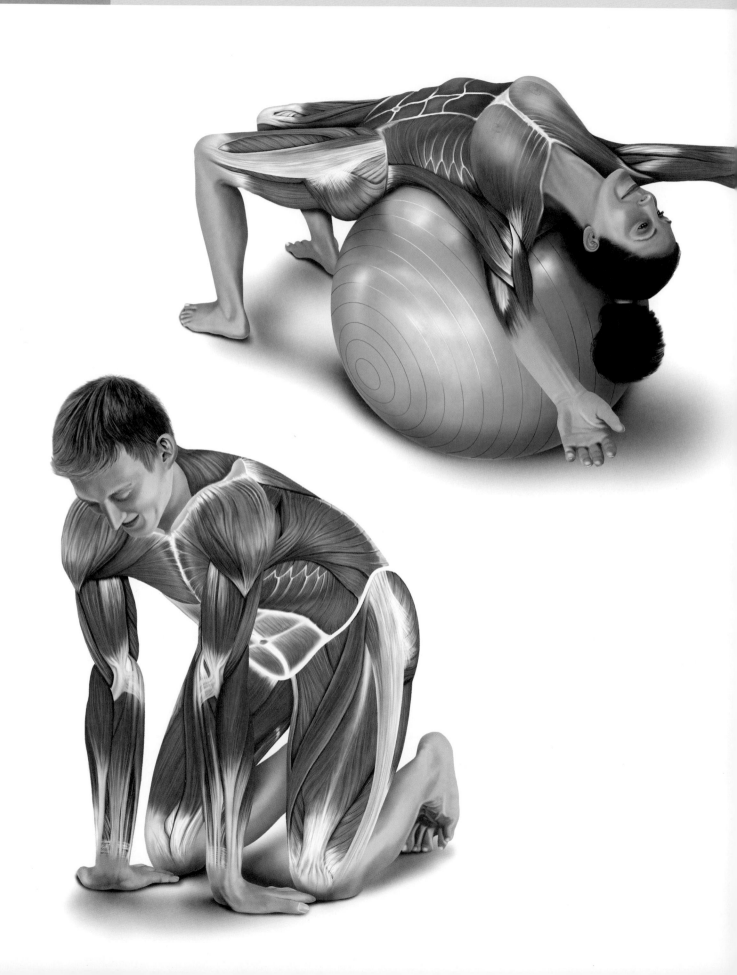

Foreword

Physical fitness is about much more than simple muscle strength and bulk. The musculoskeletal system of your body is a complex arrangement of joints, ligaments, tendons, and muscles, and the proper functioning of each of these components is essential for optimal physical performance. Gentle stretching keeps all of these elements in the very best condition, ensuring that you enjoy the fullest range of movement at your joints and that they are protected from injury. Stretching can also provide an important prelude and coda to your exercise routine.

In this book, we have collected 50 stretches that will enable you to work on all the major muscle groups in your body and to devise an appropriate stretching routine for your circumstances. Each stretch is illustrated in full color to show you not only how to perform the stretch, but also allow you to see the muscles that will get the most benefit from that particular stretch. Some muscles are deeper and are not immediately visible on the muscleman or woman, but they will benefit just as much from the stretch. Each stretch comes with specific how-to tips and special points about getting the most out of each activity.

As with all forms of physical activity, it is essential to follow the instructions for each stretch carefully, taking note of any safety points. This will minimize risk of injury and ensure that you get the greatest benefit from each stretch. If you have any instability in any of your joints (such as a previous shoulder dislocation), you should consult your medical practitioner before attempting the relevant stretch.

Incorporating stretching routines into your life is a fun way to optimize your physical fitness and achieve the greatest flexibility for your joints. It's time to take control and make stretching part of your daily routine!

Professor Ken Ashwell
Department of Anatomy,
School of Medical Sciences, Faculty of Medicine,
University of New South Wales, Sydney, Australia

How This Book Works

This book is organized into four primary sections: a full-color anatomy overview; a principles of stretching section; a full-color illustrated stretching guide, comprising the main part of the book; and a coloring workbook in which to test your anatomical knowledge.

The anatomy overview section provides detailed, anatomically correct illustrations with clear, informative labels for the various body systems and regions. Visualizing the parts of the body and their links to each other will improve your understanding of how the body works during stretching. Additional anatomy overview pages can be found at the start of each chapter in the stretching guide.

Each of the eight chapters in the stretching guide focuses on specific muscle areas—neck, shoulders, arm and forearm, trunk, back, hip and buttocks, thigh, and leg and foot. Every stretch is depicted with an anatomically correct pose. Labels identify all the important muscles—muscles being used—so you can visualize and understand exactly which muscles are activated during the exercises. This will not only increase your knowledge of anatomy; it will help to improve the effectiveness of your stretching and rehabilitation programs.

The coloring workbook chapter is a study aid that aims to facilitate your understanding of important body systems—the muscular, skeletal, and nervous systems. Color in each illustration to help memorize the location of muscles, bones, and nerves within these systems. Fill in the blank labels to test your knowledge of the names of body parts—the answers are given at the bottom of each page.

ANATOMY OVERVIEW PAGES

This section contains full-color, double-page overview spreads that give a rundown on the important parts of a particular body system. You will find additional overview pages at the beginning of each chapter in the stretching guide.

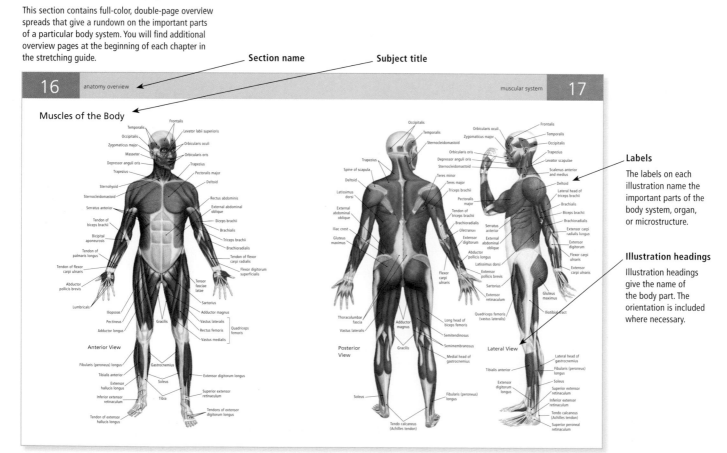

Section name

Subject title

Labels

The labels on each illustration name the important parts of the body system, organ, or microstructure.

Illustration headings

Illustration headings give the name of the body part. The orientation is included where necessary.

STRETCH PAGES

Each chapter in the stretching guide focuses on specific muscle areas. The stretches depict anatomically correct poses that identify the muscles being stretched.

Chapter name   Stretch title

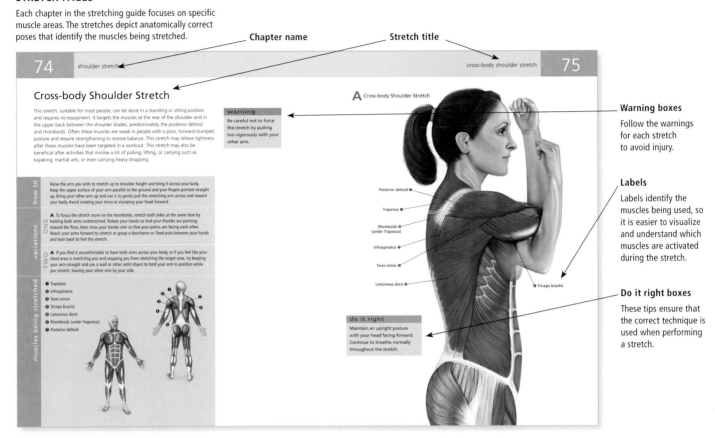

Warning boxes
Follow the warnings for each stretch to avoid injury.

Labels
Labels identify the muscles being used, so it is easier to visualize and understand which muscles are activated during the stretch.

Do it right boxes
These tips ensure that the correct technique is used when performing a stretch.

COLORING WORKBOOK PAGES

This final section contains black-and-white drawings of parts of the muscular, skeletal, and nervous systems. Color in the body parts as a memory aid.

Section name   Subject title

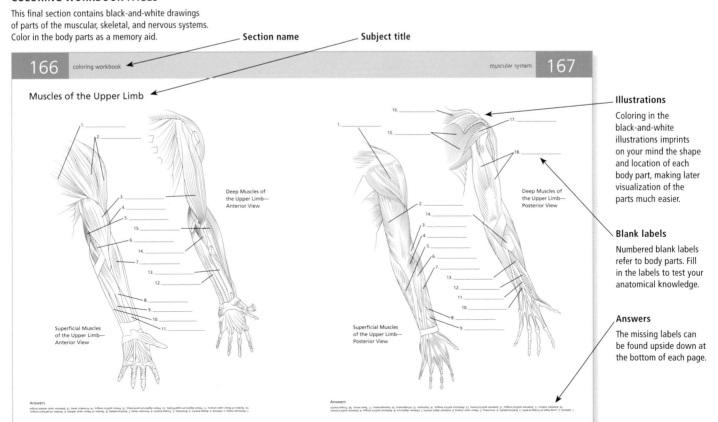

Illustrations
Coloring in the black-and-white illustrations imprints on your mind the shape and location of each body part, making later visualization of the parts much easier.

Blank labels
Numbered blank labels refer to body parts. Fill in the labels to test your anatomical knowledge.

Answers
The missing labels can be found upside down at the bottom of each page.

Anatomy Overview

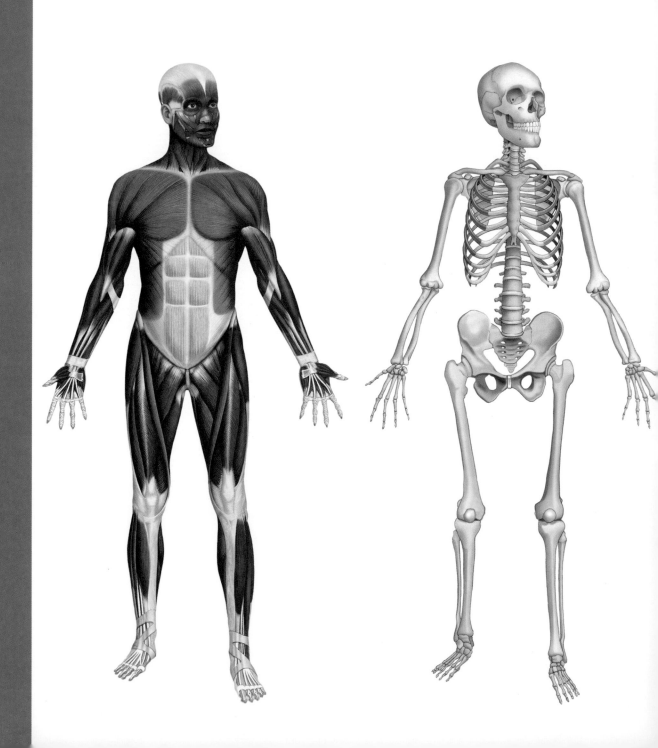

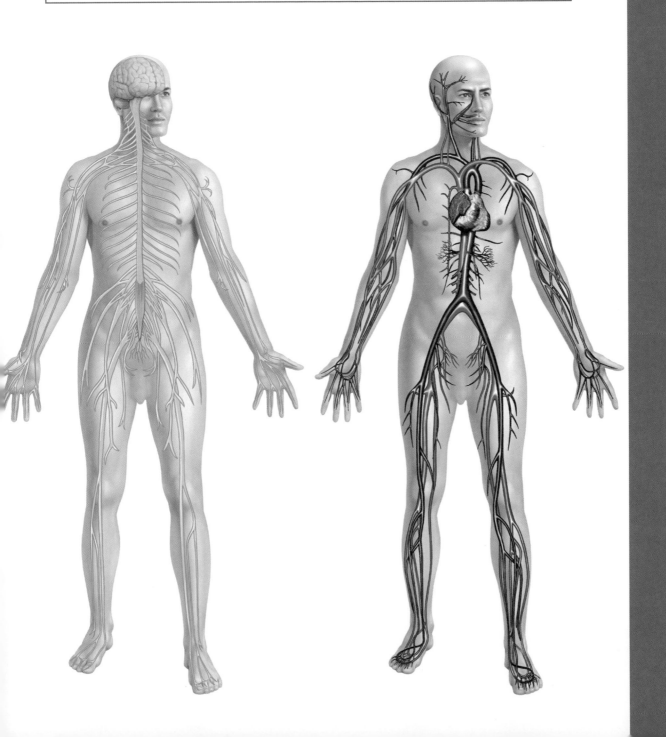

Body Regions

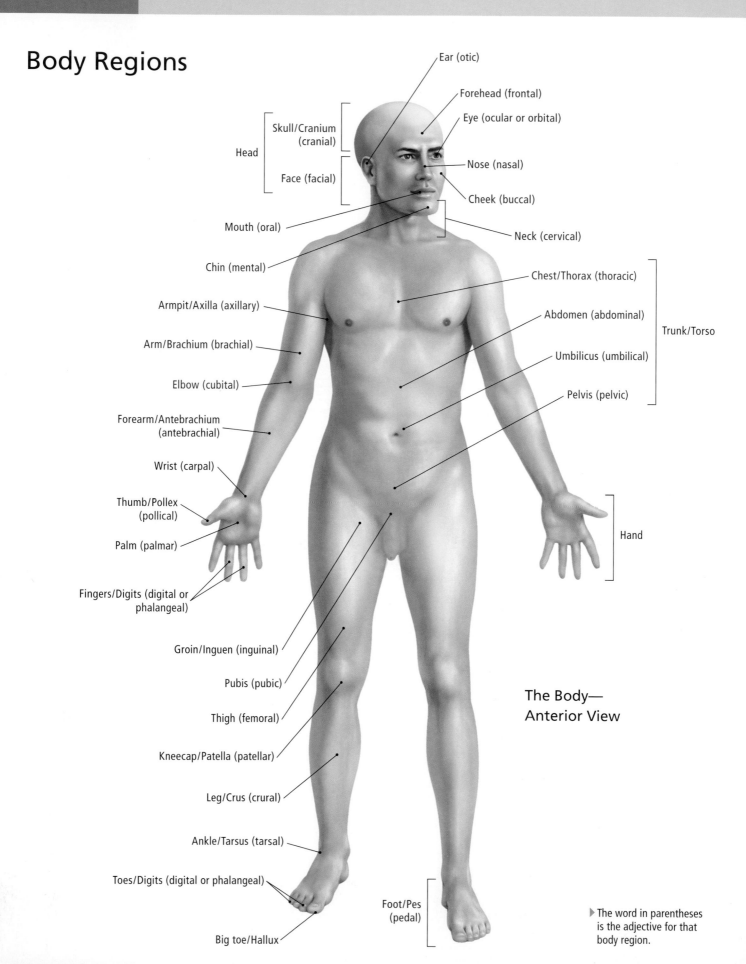

Ear (otic)

Forehead (frontal)

Eye (ocular or orbital)

Skull/Cranium (cranial)

Head

Nose (nasal)

Face (facial)

Cheek (buccal)

Mouth (oral)

Neck (cervical)

Chin (mental)

Chest/Thorax (thoracic)

Armpit/Axilla (axillary)

Abdomen (abdominal)

Arm/Brachium (brachial)

Umbilicus (umbilical)

Trunk/Torso

Elbow (cubital)

Pelvis (pelvic)

Forearm/Antebrachium (antebrachial)

Wrist (carpal)

Thumb/Pollex (pollical)

Palm (palmar)

Hand

Fingers/Digits (digital or phalangeal)

Groin/Inguen (inguinal)

Pubis (pubic)

The Body—Anterior View

Thigh (femoral)

Kneecap/Patella (patellar)

Leg/Crus (crural)

Ankle/Tarsus (tarsal)

Toes/Digits (digital or phalangeal)

Foot/Pes (pedal)

Big toe/Hallux

▶ The word in parentheses is the adjective for that body region.

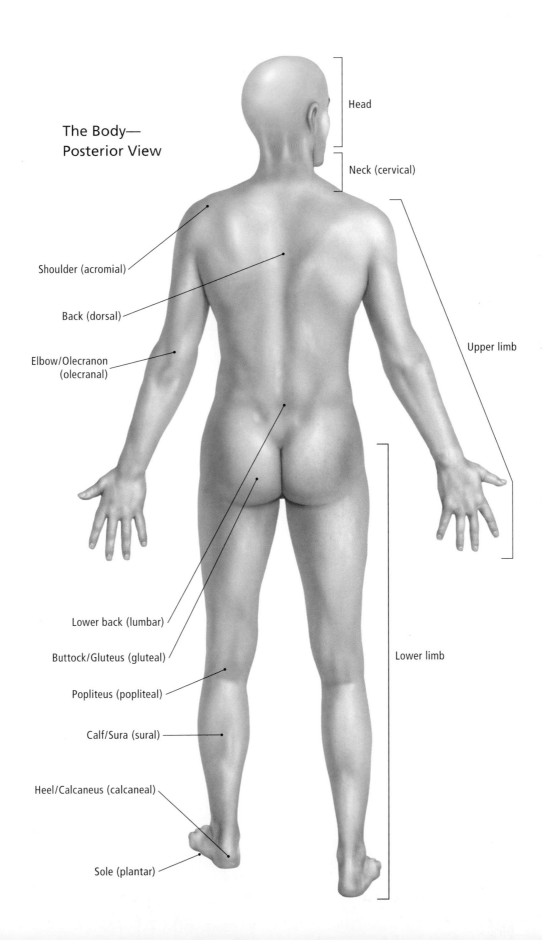

The Body—
Posterior View

Head

Neck (cervical)

Shoulder (acromial)

Back (dorsal)

Elbow/Olecranon
(olecranal)

Upper limb

Lower back (lumbar)

Buttock/Gluteus (gluteal)

Popliteus (popliteal)

Lower limb

Calf/Sura (sural)

Heel/Calcaneus (calcaneal)

Sole (plantar)

Muscles of the Body

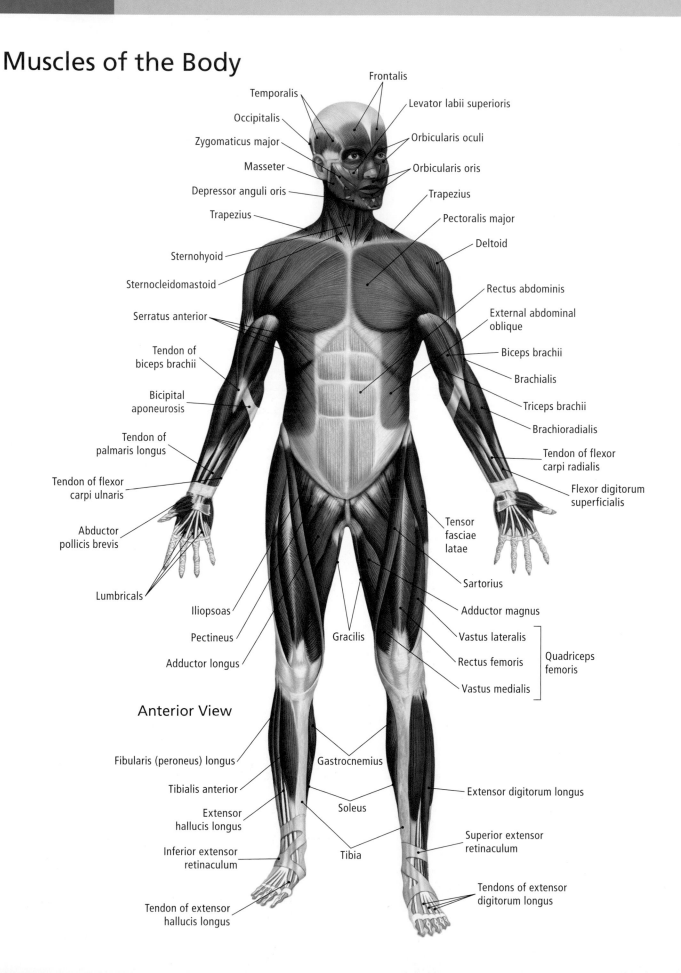

Frontalis

Temporalis

Levator labii superioris

Occipitalis

Orbicularis oculi

Zygomaticus major

Orbicularis oris

Masseter

Trapezius

Depressor anguli oris

Pectoralis major

Trapezius

Deltoid

Sternohyoid

Rectus abdominis

Sternocleidomastoid

External abdominal oblique

Serratus anterior

Biceps brachii

Tendon of biceps brachii

Brachialis

Triceps brachii

Bicipital aponeurosis

Brachioradialis

Tendon of palmaris longus

Tendon of flexor carpi radialis

Tendon of flexor carpi ulnaris

Flexor digitorum superficialis

Abductor pollicis brevis

Tensor fasciae latae

Sartorius

Lumbricals

Adductor magnus

Iliopsoas

Vastus lateralis

Quadriceps femoris

Pectineus

Rectus femoris

Gracilis

Adductor longus

Vastus medialis

Anterior View

Fibularis (peroneus) longus

Gastrocnemius

Tibialis anterior

Extensor digitorum longus

Extensor hallucis longus

Soleus

Superior extensor retinaculum

Inferior extensor retinaculum

Tibia

Tendon of extensor hallucis longus

Tendons of extensor digitorum longus

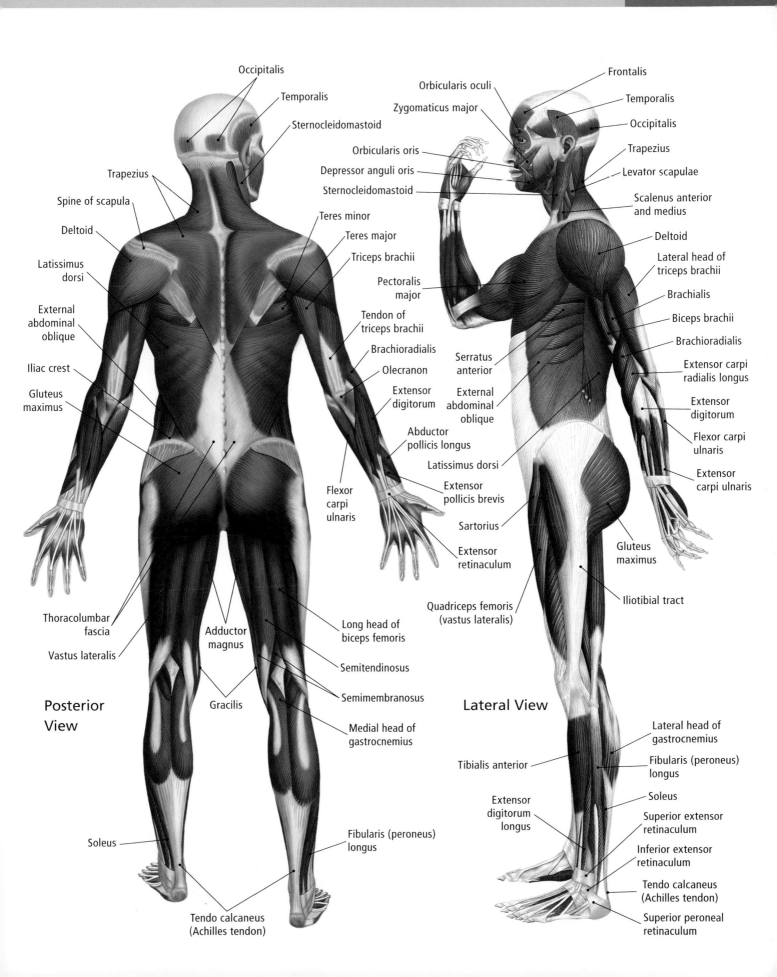

Posterior View

Occipitalis
Temporalis
Sternocleidomastoid
Trapezius
Spine of scapula
Deltoid
Latissimus dorsi
External abdominal oblique
Iliac crest
Gluteus maximus
Teres minor
Teres major
Triceps brachii
Pectoralis major
Tendon of triceps brachii
Brachioradialis
Olecranon
Extensor digitorum
Flexor carpi ulnaris
Thoracolumbar fascia
Vastus lateralis
Adductor magnus
Gracilis
Long head of biceps femoris
Semitendinosus
Semimembranosus
Medial head of gastrocnemius
Soleus
Fibularis (peroneus) longus
Tendo calcaneus (Achilles tendon)

Lateral View

Orbicularis oculi
Zygomaticus major
Orbicularis oris
Depressor anguli oris
Sternocleidomastoid
Serratus anterior
External abdominal oblique
Latissimus dorsi
Extensor pollicis brevis
Sartorius
Extensor retinaculum
Quadriceps femoris (vastus lateralis)
Frontalis
Temporalis
Occipitalis
Trapezius
Levator scapulae
Scalenus anterior and medius
Deltoid
Lateral head of triceps brachii
Brachialis
Biceps brachii
Brachioradialis
Extensor carpi radialis longus
Extensor digitorum
Flexor carpi ulnaris
Extensor carpi ulnaris
Gluteus maximus
Iliotibial tract
Abductor pollicis longus
Tibialis anterior
Extensor digitorum longus
Lateral head of gastrocnemius
Fibularis (peroneus) longus
Soleus
Superior extensor retinaculum
Inferior extensor retinaculum
Tendo calcaneus (Achilles tendon)
Superior peroneal retinaculum

Bones of the Body

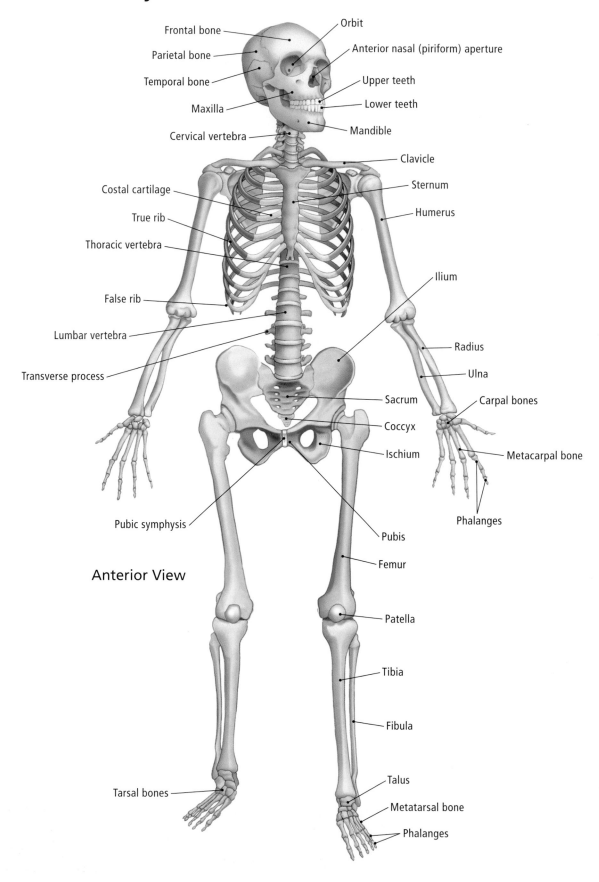

Anterior View

Frontal bone
Parietal bone
Temporal bone
Maxilla
Cervical vertebra
Orbit
Anterior nasal (piriform) aperture
Upper teeth
Lower teeth
Mandible
Clavicle
Costal cartilage
True rib
Thoracic vertebra
False rib
Lumbar vertebra
Transverse process
Sternum
Humerus
Ilium
Radius
Ulna
Carpal bones
Metacarpal bone
Phalanges
Sacrum
Coccyx
Ischium
Pubic symphysis
Pubis
Femur
Patella
Tibia
Fibula
Talus
Metatarsal bone
Phalanges
Tarsal bones

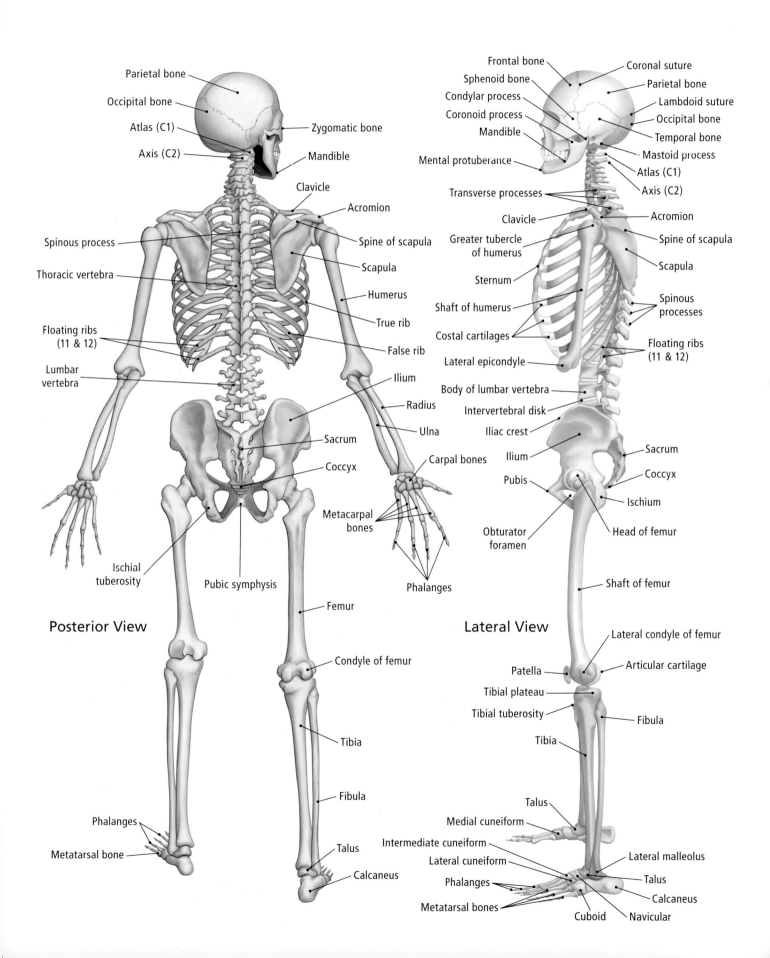

Parietal bone

Occipital bone

Atlas (C1)

Axis (C2)

Zygomatic bone

Mandible

Clavicle

Acromion

Spine of scapula

Scapula

Humerus

True rib

False rib

Spinous process

Thoracic vertebra

Floating ribs
(11 & 12)

Lumbar
vertebra

Ilium

Radius

Ulna

Sacrum

Coccyx

Carpal bones

Metacarpal
bones

Phalanges

Ischial
tuberosity

Pubic symphysis

Femur

Condyle of femur

Tibia

Fibula

Talus

Calcaneus

Phalanges

Metatarsal bone

Posterior View

Frontal bone

Sphenoid bone

Condylar process

Coronoid process

Mandible

Mental protuberance

Coronal suture

Parietal bone

Lambdoid suture

Occipital bone

Temporal bone

Mastoid process

Atlas (C1)

Axis (C2)

Transverse processes

Clavicle

Greater tubercle
of humerus

Sternum

Shaft of humerus

Costal cartilages

Lateral epicondyle

Body of lumbar vertebra

Intervertebral disk

Iliac crest

Ilium

Pubis

Obturator
foramen

Acromion

Spine of scapula

Scapula

Spinous
processes

Floating ribs
(11 & 12)

Sacrum

Coccyx

Ischium

Head of femur

Shaft of femur

Lateral condyle of femur

Articular cartilage

Patella

Tibial plateau

Tibial tuberosity

Fibula

Tibia

Talus

Medial cuneiform

Intermediate cuneiform

Lateral cuneiform

Phalanges

Metatarsal bones

Lateral malleolus

Talus

Calcaneus

Cuboid

Navicular

Lateral View

Nervous System

Nervous System—Anterior View

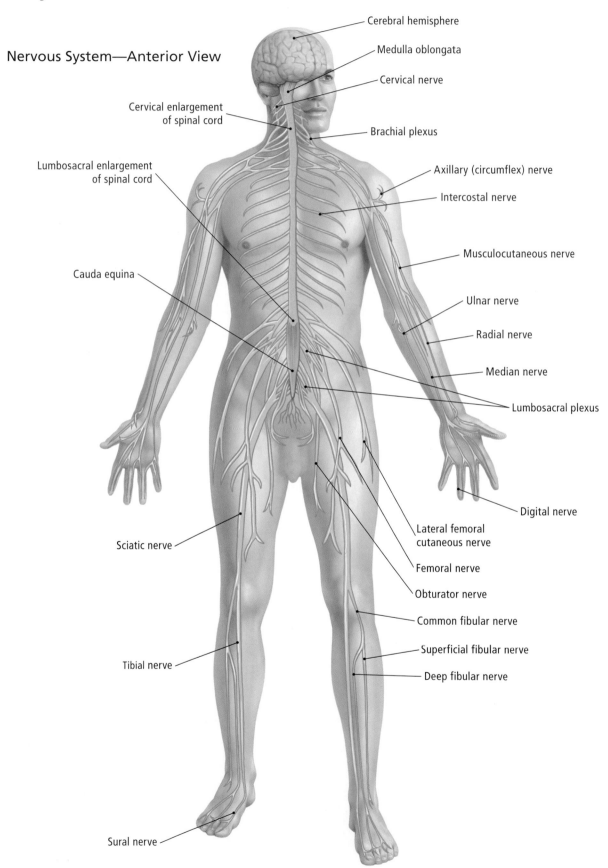

Cerebral hemisphere

Medulla oblongata

Cervical nerve

Cervical enlargement
of spinal cord

Brachial plexus

Lumbosacral enlargement
of spinal cord

Axillary (circumflex) nerve

Intercostal nerve

Musculocutaneous nerve

Cauda equina

Ulnar nerve

Radial nerve

Median nerve

Lumbosacral plexus

Digital nerve

Lateral femoral
cutaneous nerve

Sciatic nerve

Femoral nerve

Obturator nerve

Common fibular nerve

Superficial fibular nerve

Tibial nerve

Deep fibular nerve

Sural nerve

Circulatory System

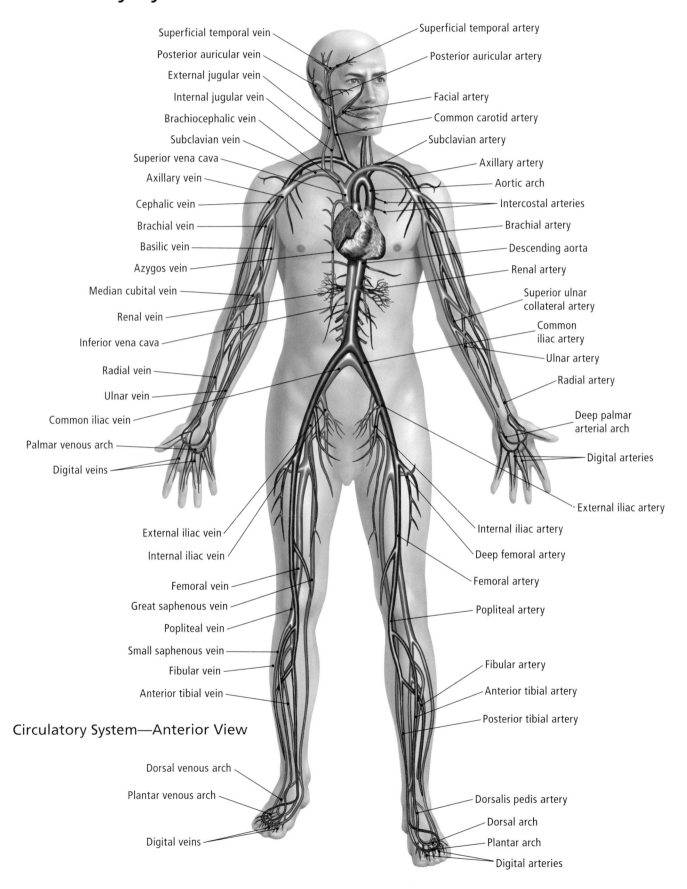

Superficial temporal vein
Posterior auricular vein
External jugular vein
Internal jugular vein
Brachiocephalic vein
Subclavian vein
Superior vena cava
Axillary vein
Cephalic vein
Brachial vein
Basilic vein
Azygos vein
Median cubital vein
Renal vein
Inferior vena cava
Radial vein
Ulnar vein
Common iliac vein
Palmar venous arch
Digital veins

Superficial temporal artery
Posterior auricular artery
Facial artery
Common carotid artery
Subclavian artery
Axillary artery
Aortic arch
Intercostal arteries
Brachial artery
Descending aorta
Renal artery
Superior ulnar collateral artery
Common iliac artery
Ulnar artery
Radial artery
Deep palmar arterial arch
Digital arteries
External iliac artery
Internal iliac artery
Deep femoral artery
Femoral artery
Popliteal artery
Fibular artery
Anterior tibial artery
Posterior tibial artery

External iliac vein
Internal iliac vein
Femoral vein
Great saphenous vein
Popliteal vein
Small saphenous vein
Fibular vein
Anterior tibial vein

Circulatory System—Anterior View

Dorsal venous arch
Plantar venous arch
Digital veins

Dorsalis pedis artery
Dorsal arch
Plantar arch
Digital arteries

Movements of the Body

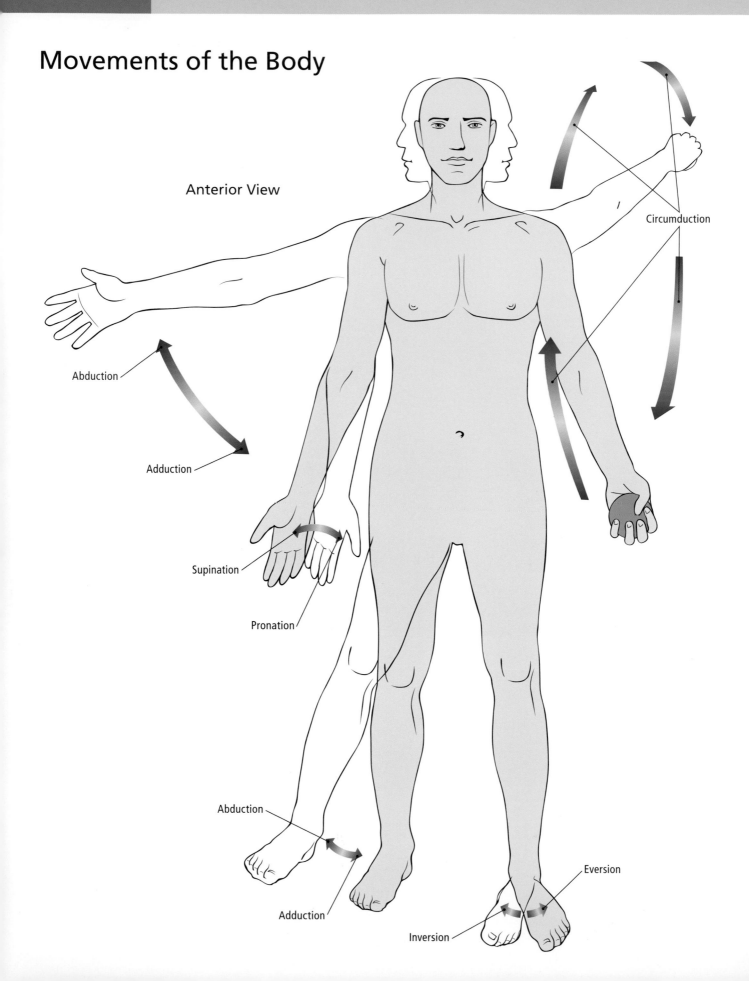

Anterior View

Abduction

Adduction

Supination

Pronation

Abduction

Adduction

Circumduction

Eversion

Inversion

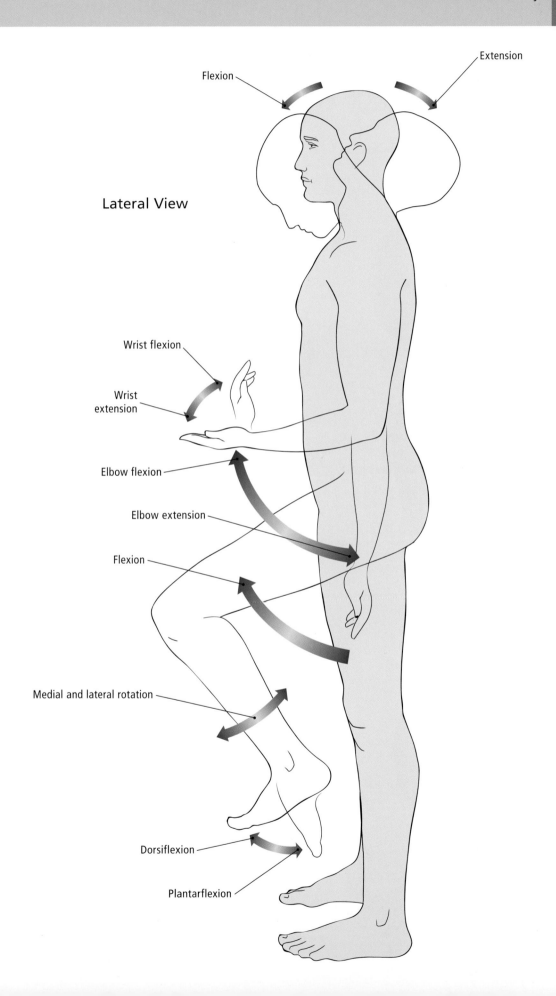

Flexion

Extension

Lateral View

Wrist flexion

Wrist extension

Elbow flexion

Elbow extension

Flexion

Medial and lateral rotation

Dorsiflexion

Plantarflexion

The Principles of Stretching

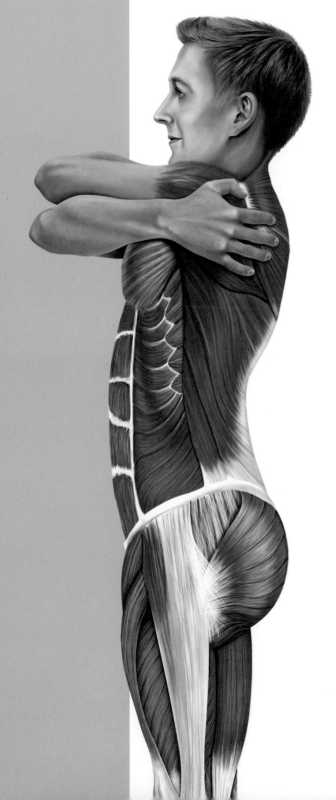

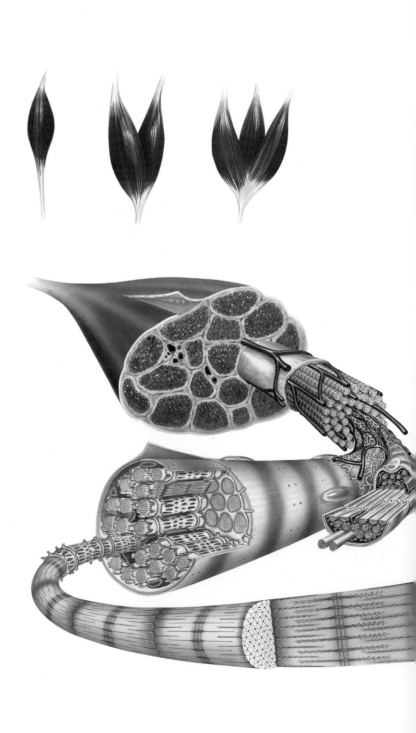

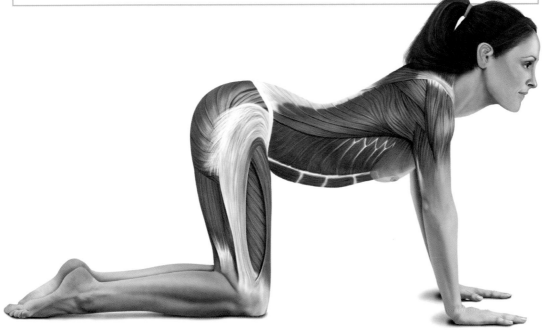

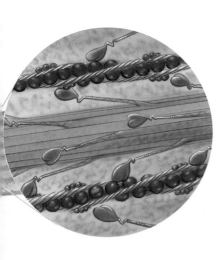

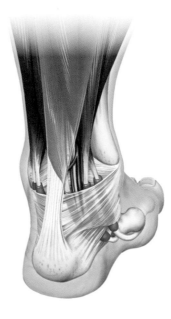

What is Stretching

Stretching is a form of exercise. When you stretch, you are deliberately lengthening soft tissue structures within your body. Stretching is usually focused on muscle tissue, but other types of soft tissue within the body may be targeted including tendons, ligaments, nerves, and skin. There are many reasons to stretch. These include relaxing tense muscles, restoring muscle length following injury or periods of immobilization, improving flexibility, improving posture, as part of a warm-up before more vigorous exercise, improving performance in certain activities, or simply because it feels good. With some careful consideration, stretching can be a very effective way of achieving your exercise goals. Don't get too hung up about it, though. You don't need to be as flexible as a contortionist to get the benefits of stretching!

Types of Tissue

Stretching acts on many different tissues in the body. Primarily, stretching acts on skeletal muscle, but it may also have effects on ligaments, tendons, loose connective tissue, and even nerve (neural) tissue.

Loose connective tissue

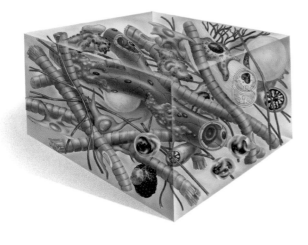

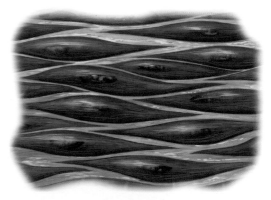

Muscle tissue

Smooth muscle

Skeletal muscle

Cardiac muscle

Stretching is natural. Humans (and animals) often stretch without giving it much conscious thought. A good example of this is yawning and stretching as you wake from sleep. We do this to relieve stiffened muscles and joints that have been held in one position for a long time.

Deliberate stretching routines can take many different forms and can aid all areas of the body. Static stretches are the best-known and most commonly performed type. They are very effective for enhancing or improving range of motion and for aiding relaxation. Dynamic stretching is a key part of many sports or exercise warm-ups. Other types of stretching are typically used by health professionals to assist in injury rehabilitation or to aid sports performance. These stretch techniques include partner assisted, mechanically assisted, and those incorporating contract-relax techniques or joint or nerve mobilization techniques.

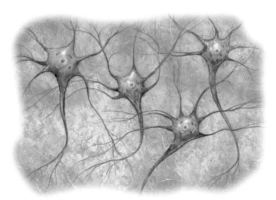

Neural tissue

Tendon tissue (relaxed)

Ligament tissue

What is Stretching

Before looking at the benefits of stretching, it is worth clearing up some common misconceptions. Stretching is often recommended for prevention of injury and improved athletic performance. Unfortunately, things are not as simple as that. While there are times when stretching may reduce the risk of injury or enable performance of certain tasks, don't believe that all stretching is always beneficial. Stretching can in fact be detrimental to performance of some activities and can cause injury if performed incorrectly or inappropriately. Therefore, it is always important to be sure of why, how, and when you should be stretching.

Although there is no compelling evidence that stretching inevitably helps to reduce injury, it is still commonly prescribed and performed in clinical practice when perceived deficiencies or imbalances are noted. If you feel that you have a particular injury risk, then a physical screening by a qualified health professional may identify areas of concern, or reveal a less than optimum range of motion, and stretching may be of some benefit in these situations.

Types of Tissue
Many different tissues of the body may be affected by stretching.

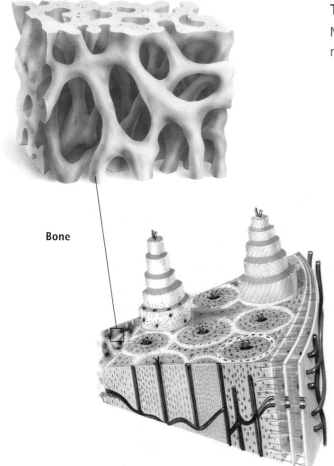

Bone

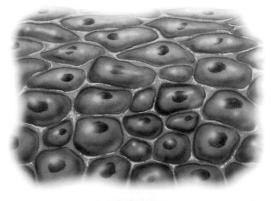

Epithelial tissue

Static stretches for warm-up are generally not recommended. Stretching and holding a muscle in a lengthened state will do nothing to make it warmer! It is a good idea to warm muscles up before exercising. They will move better, but do this with a walk, slow jog, or dynamic stretching routine.

Many people stretch in the hope of preventing delayed onset muscle soreness (DOMS) after exercise. Again, this commonly held belief is simply not true! There is no evidence that stretching helps reduce muscle soreness, and many scientific studies have clearly debunked this idea.

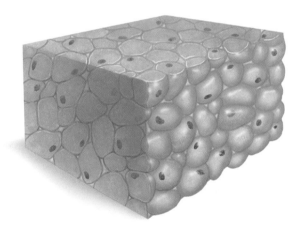

Adipose tissue

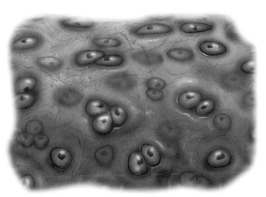

Elastic cartilage

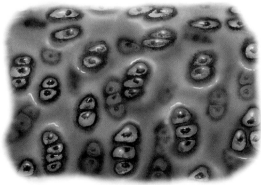

Hyaline cartilage

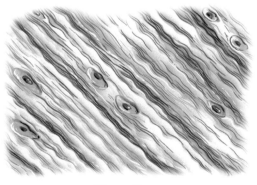

Fibrocartilage

Cartilage tissue

What is Stretching

Stretching is likely to improve the athletic performance of gymnasts and divers who need to get into positions that most of us don't need to. Similarly, there is good evidence that swimmers can swim faster with less power output if they are flexible enough to achieve an adequate streamline position. But there is equally good evidence that sprint performance is actually reduced by performing repeated static stretches. So, if you think stretching is going to help your sports performance, be aware of what you are doing and why.

Stretching, however, feels good, and most of us feel the need to do it from time to time. Stretching certainly does something, just not what people often believe it is doing. Stretching helps people to move more freely. Muscles and joints feel stiff if they are not moved, so regular movement makes us feel better. Following injury, when muscles and joints have adapted to a shortened state, stretching is a useful part of the rehabilitation process.

Stretching for Exercise
Stretching may be of benefit to gymnasts and any athlete who needs to adopt unusual body positions. Stretching also feels good and may help improve your state of mind.

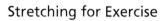 Standing Outer Hip stretch, p.132

Keep in mind that some people simply respond better to stretching than others; it is in their genes. So, if you are determined to stretch, but still can't touch your toes, don't despair; just realize that it probably isn't going to happen. You won't be an Olympic gymnast, but there are plenty of benefits completely separate from muscle length. There is even evidence that stretching is good for your heart! So stretch because it feels good and stretch because you want to, not because someone told you "you have to."

▶ Cow Stretch, p.116

Muscles, Joints, and Nerves

When you stretch, tension is applied not just to muscles, but also to the tendons by which muscles insert into bones and the ligaments that maintain the stability of your joints. Stretching also affects the nervous system, by stimulating the reflex pathways that naturally control muscle tone.

Muscle Structure

Muscle cells are grouped together in bundles surrounded by connective tissue. Each muscle cell contains a nucleus (for cellular control) situated at the edge of the cell, directly under the cell membrane (sarcolemma). About 80 percent of the volume of each muscle cell is made up of fiber bundles called myofibrils surrounded by small cellular structures called mitochondria that produce energy. The myofibrils are made up of two major types of filaments: thin filaments that contain actin and thick filaments composed of myosin.

Muscles produce movement by active shortening. This action is stimulated by nerve impulses from motorneurons in the spinal cord that reach the motor end-plate and activate the release of calcium inside the muscle cell. The release of calcium triggers sliding of muscle filaments inside the muscle cells, a process that requires oxygen and chemical energy produced from food. It is the sliding of actin filaments relative to the myosin filaments that produces the actual muscle contraction. During contraction, the muscle may shorten to about one-third of its original length. Conversely, the process of stretching lengthens the muscle and requires that the muscle filaments slide in the opposite direction to that seen in contraction. This stretching allows muscle cells to reach their full potential length.

Neuron

Neurons are specialized cells found in the nervous system that conduct nerve impulses. Each neuron has three main parts: the cell body, the branching projections (dendrites) that carry impulses to the cell body, and one elongated projection (the axon) that conveys impulses away from the cell body.

Synaptic knob

Axon terminal

Axon

Myelin sheath

Cell body

Golgi apparatus

Nuclear membrane

Nucleolus

Mitochondrion

Dendrite

The attachments of muscles to bone and other tissue may be in the form of tendons (rounded connective tissue bands) or sheets of tough membrane called aponeuroses. These structures will also be placed under tension by the stretches described in this book. Tendons will not extend in the short term, but may undergo some minor lengthening when stretching is applied over a long period of time. More importantly, the long-term application of tension during stretching will increase the tensile strength of tendons and aponeuroses.

Muscle Fiber—Microstructure

A muscle fiber or cell consists of myofibrils (contractile proteins) surrounded by nuclei just under the cell membrane (sarcolemma). Contraction of the muscle is produced by the interaction of myosin heads with the protein actin.

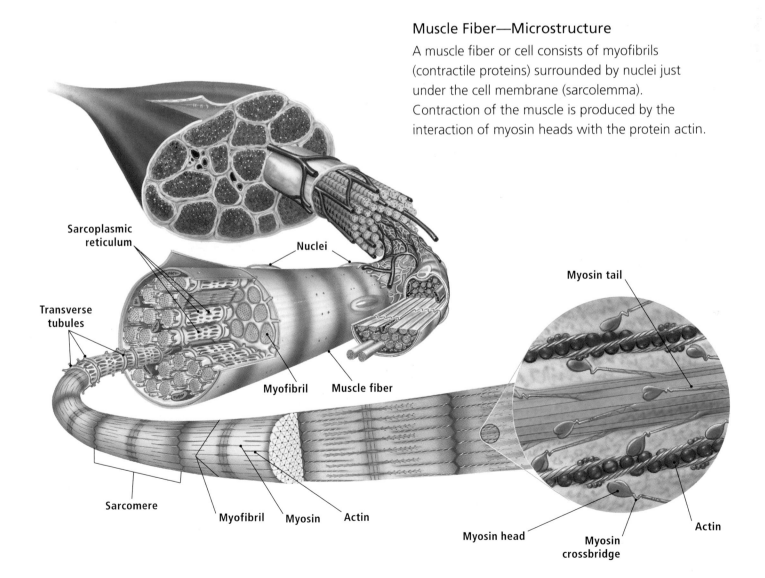

Sarcoplasmic reticulum

Nuclei

Transverse tubules

Myofibril

Muscle fiber

Myosin tail

Sarcomere

Myofibril

Myosin

Actin

Myosin head

Myosin crossbridge

Actin

Muscles, Joints, and Nerves

Muscle Shapes

Muscles that attach to the skeleton (skeletal muscle) come in many different shapes and sizes, and particular muscle shapes can influence the effect that various stretching positions have on muscles. The arrangement of muscle fibers relative to the tendon or aponeurosis of a muscle determines the relative importance of muscle power versus range of movement for the particular muscle. This also has implications for the way that the muscle tolerates being stretched.

For example, long thin straplike muscles like the sartorius or gracilis muscles of the thigh have many muscle cells arranged in series (muscle fibers arranged in rows one after another), so the muscle may shorten by as much as 50 percent during contraction. This also means that the muscle may tolerate pronounced stretching without muscle damage. Fusiform muscles, like the biceps brachii, are also reasonably tolerant of stretching, because they have muscles fibers arranged mainly in series.

Muscle Shapes

Muscles are classified based on their general shape—some muscles have mainly parallel fibers and others have oblique fibers. The shape and arrangement of muscle fibers reflect the function of the muscle (e.g., muscle fibers that support organs are crisscrossed).

Unipennate   **Bipennate**   **Multipennate**   **Spiral**   **Spiral**   **Radial**

On the other hand, pennate muscles (like the vastus medialis part of quadriceps femoris, or the long flexors of the digits) have many muscle fibers arranged in parallel, as they insert into the side of a long tendon, producing the feathered appearance that gives these muscles their name (from penna, the Latin for feather). This allows a lot of muscle power to be applied to the tendon, but does not allow as much lengthening of the muscle. This also means that pennate muscles tolerate stretching less well than strap or fusiform muscles.

Some muscles have multiple parts, each of which may be stretched by different movements. An example of this type is the deltoid muscle, which has anterior, lateral, and posterior parts, each of which is stretched by a different movement of the shoulder. Flexion, or forward movement, of the shoulder will stretch the posterior parts, whereas extension, or backward movement, will stretch the anterior fibers.

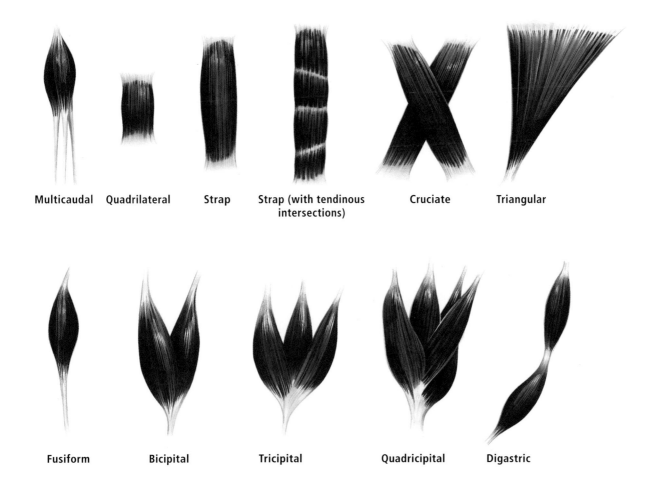

Multicaudal **Quadrilateral** **Strap** **Strap (with tendinous intersections)** **Cruciate** **Triangular**

Fusiform **Bicipital** **Tricipital** **Quadricipital** **Digastric**

Muscles, Joints, and Nerves

Joint Structure

Stretching puts many joints through their full range of movement and applies tension to the capsule around the joint space. Such stretches gently extend the capsule and ligaments around the joint in ways that may not be naturally encountered during daily activity, but nevertheless assist joint mobility for exercise.

The joints most affected by stretching are the synovial joints, meaning they contain a joint space filled with lubricating synovial fluid, surrounded by a synovial membrane and protective joint capsule. The surfaces of the bone in synovial joints are covered with hyaline cartilage that is compressed during weight bearing but returns to its original shape when the compressive force is removed. The synovial membrane makes the synovial fluid that lubricates movement of the cartilage surfaces against each other.

Synovial joints

Synovial joints are the most mobile joints in the body. The shape of articular cartilage surfaces in a synovial joint and the way they fit together determine the range and direction of the joint's movement.

Saddle joint

A saddle joint allows movement in two planes, but not rotation around the long axis of the bone.

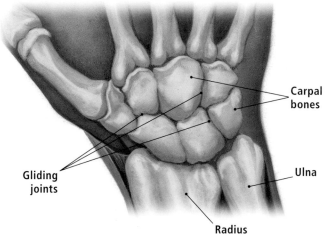

Carpal bones

Gliding joints

Ulna

Radius

Gliding joint

Aided by synovial fluid, the bones in the joint slide across each other in a limited movement.

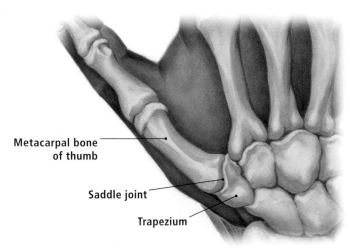

Metacarpal bone of thumb

Saddle joint

Trapezium

Synovial joint stability is due to a combination of factors, including joint shape, the arrangement of the joint capsule fibers, the alignment of associated ligaments, and the strength of muscles around the joint (e.g., the rotator cuff muscles around the shoulder joint). Joints where adjacent cartilage surfaces have a high area of contact (such as the hip joint) are particularly mechanically stable. Regular stretching has no bearing on this stability, but it can assist mobility by applying tension to the joint capsule and associated ligaments around the joint, producing gradual lengthening of their connective tissue fibers. Exercise and gentle stretching also strengthen the muscles around the joint. Rather than producing joint laxity, stretching allows the joint to develop an optimal range of movement while maintaining stability through protective muscle activity.

Many of the stretches in this book act on limb joints with a wide range of movement (e.g., the knee and shoulder joints). Other stretches produce tiny increases in mobility of a series of small joints (e.g., the joints of the vertebral column).

Pivot joint

The joint between the first and second cervical vertebrae rotates—this is known as a pivot joint.

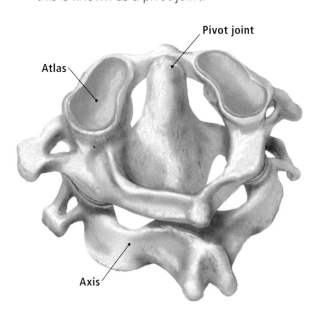

Pivot joint

Atlas

Axis

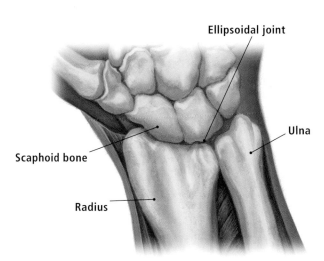

Ellipsoidal joint

Scaphoid bone

Radius

Ulna

Ellipsoidal joint

This is a joint structure that allows movement in two directions, such as that which takes place at the wrist joint.

Muscles, Joints, and Nerves

Stretching and Nerves

Apart from applying tension to muscles, tendons, and ligaments, stretching stimulates the natural reflex mechanisms that control the tone and length of muscles. Muscles and tendons have sensory structures embedded inside the muscle itself (the muscle spindles that detect lengthening of the muscle), or within the tendon (Golgi tendon organs that detect changes in muscle force). Information from these sensory structures is carried back to the spinal cord by axons of the dorsal root ganglion cells that make up peripheral nerves. Once in the spinal cord, the information is used to regulate muscle function during active movement.

Nerve cells called motorneurons, which are found in the anterior horn of spinal cord gray matter, produce movement of muscles by sending impulses down their axons in peripheral nerves to the muscle cell.

Reflexes

Reflexes are quick and automatic responses to stimuli, whereby a nerve impulse travels to a nerve center in the spinal cord. The nerve center then sends a message outward to a muscle or gland to effect a response without the person being consciously aware of it.

Spinal nerve sends signal along peripheral nerves to muscle cells

Muscle is activated by signal from motor nerve cells

e

d

c

Spinal cord (central nervous system) processes information

Receptors send message along nerve fibers to spinal cord

b

a

Stimulus is registered by sensory receptors

The reflex pathway that is most affected by muscle stretching is the deep tendon or stretch reflex. This is a simple pathway that uses information from the muscle spindles, which detect changes in the length of the muscle, to regulate muscle tone. The reflex involves only two nerve cells: a sensory nerve cell (dorsal root ganglion cell), which has a process that runs into the anterior horn of the spinal cord to contact the other nerve cell of the pathway, and the motorneuron that will in turn drive the muscle. Sudden stretching of the muscle (e.g., by tapping the tendon) excites a series of nerve impulses that run up the sensory nerve cell into the spinal cord and activate the motorneuron to produce a jerky movement. The activity of the stretch reflex is usually dampened down by the influence of nerve pathways from the brainstem, assisting the mobility of muscles and joints during exercise.

Gentle stretching of major muscles allows the descending pathways from the brainstem to regulate muscle tone over a wide range of limb movements, keeping joints mobile and muscles supple, while still maintaining muscle strength.

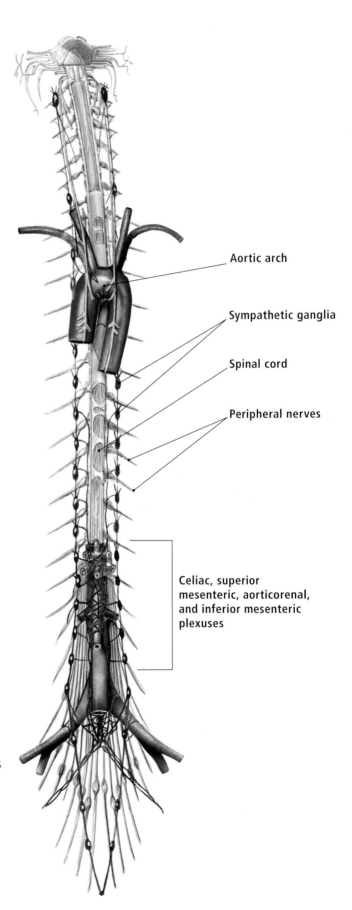

Aortic arch

Sympathetic ganglia

Spinal cord

Peripheral nerves

Celiac, superior mesenteric, aorticorenal, and inferior mesenteric plexuses

Spinal Cord—Anterior View

The spinal cord is an important center for reflex activity. It contains the point of contact between sensory nerve cells and motorneurons, and projects commands to muscles by peripheral nerves.

Different Types of Stretches

There are a great many different stretch types and many different stretches that can be performed for any one body part or muscle group. When considering what type of stretches are best for you, it is important to think about why you are stretching and what your goals are, as this will dictate the types of stretches you should perform. What follows is a list of different categories of stretches and what they might be used for.

Static

Static stretching means to stretch with a constant or continued hold of the stretch position. Static stretches can be performed many times with relatively short hold periods or fewer times with longer hold periods. There is some debate as to which is the most effective method; however, recent research suggests that there is no difference if the total stretch time is the same (i.e., 6 x 10 second holds or 2 x 30 second holds). Static stretching is the most common form of stretching, and is perfect for improving muscle length in tight muscles, restoring muscle length after injury or immobilization, relaxation, and improving stretch pain tolerance.

Static Stretching

Static stretching involves a constant or sustained hold of the stretch position. It will improve muscle length if the muscle is tight or has been immobilized after injury.

▶ Kneeling Forearm stretch, p.90

Dynamic

Dynamic stretching, often referred to as ballistic stretching, is the best type of stretching to perform as part of a warm-up before exercise or sporting performance. It is arguable whether this is technically stretching, as there is no hold component, but it is included here, as it is commonly referred to in this way. Dynamic stretching involves slow to moderate movement through certain ranges of motion that are specific or similar to the sporting movement that is to follow. This type of stretching is not intended to lengthen muscles, but to help prepare the mind and body for the upcoming performance, to warm up the muscles, and to increase blood flow to the working muscles.

▶ Side Lunge stretch, p.146

Dynamic Stretching

Dynamic stretching is mainly used to prepare a muscle or set of muscles for movements during particular sporting activities. It involves moving into and out of the stretch in a smooth movement, but does not require a sustained hold in the stretch position.

Different Types of Stretches

Serial casting

This fairly rare stretching type is sometimes used following surgical repair of torn muscles or tendons. Serial casting involves placing the targeted muscle at a certain length using a plaster or fiberglass cast. After a period, the cast is removed, the joint is moved to a different angle (hence changing the muscle length), and recast. This can be repeated as many times as required. For example, following surgery on the Achilles tendon, the ankle may be cast at 130 degrees plantar flexion for two weeks, then changed to 110 degrees for two weeks, and then 90 degrees for two weeks.

PNF

Proprioceptive neuromuscular facilitation (PNF) stretching (also known as contract–relax stretching) is particularly effective for stretching stubborn muscles or relaxing very tight muscles. This can be done individually or as a partner-assisted stretch. To perform a PNF stretch, the muscle is taken to its end-of-range position and held. Then the muscle is contracted, or tensioned back, against the stretch direction without moving. After holding the contraction for a few seconds, the muscle is relaxed and then stretched a little further. Repeat as required. It is quite surprising how much farther a muscle can be stretched from its original position with this technique.

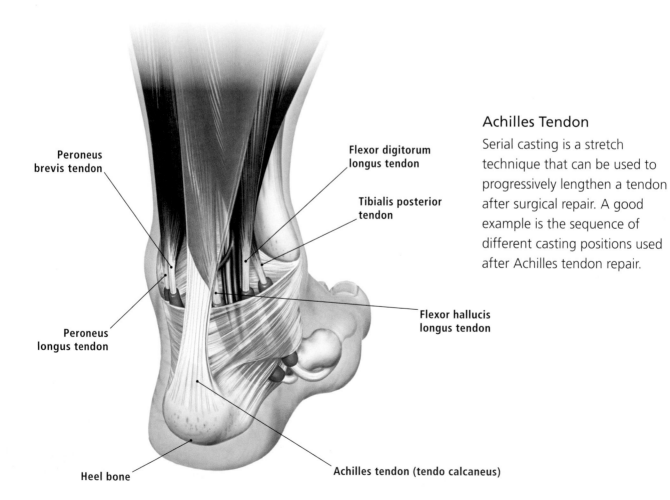

Peroneus brevis tendon

Flexor digitorum longus tendon

Tibialis posterior tendon

Achilles Tendon

Serial casting is a stretch technique that can be used to progressively lengthen a tendon after surgical repair. A good example is the sequence of different casting positions used after Achilles tendon repair.

Peroneus longus tendon

Flexor hallucis longus tendon

Heel bone

Achilles tendon (tendo calcaneus)

Partner assisted

As the name suggests, this type of stretching requires two people working together. One is the "stretchee" and one is the "stretcher." The stretcher helps his or her partner into the stretch and then holds the limb in the stretch position. As the stretchee feels the muscle relax, he or she can ask the stretcher to push further, or they can work together to perform a PNF stretch.

Partner Assisted

Partner assisted stretches may be used to achieve positions that are difficult for the individual to enter on his or her own. Partner assistance is also beneficial when the "stretchee" has problems with balance or stability.

▶ Kneeling Back Rotation stretch, p.118

Different Types of Stretches

Mechanical

Mechanical stretching (using machines such as traction units) is generally used in an injury rehabilitation setting and is often used to stretch the muscles, ligaments, and joints of the spine. This can assist injured discs, "pinched" nerves, or help relieve pain in compressed joints.

Neural or joint mobilization

One very significant variation to muscle-targeted stretching is stretching that targets neural (nerve) structures or joints. This often forms an important part of rehabilitation after injury or surgery, when accumulated scar tissue can stop the natural movement of the nerve or joint. These stretches are never static. Mobilization involves back-and-forth motions to enable free and easy movement of the nerve within its nerve sheath or the joint and joint capsule.

Mechanical

Mechanical stretching may be of benefit in relieving the pain associated with herniated intervertebral disks or compressed joints. Herniated disks may compress spinal nerves and cause referred pain to the leg, so relieving that compression will ease pressure on the nerve and the referred pain.

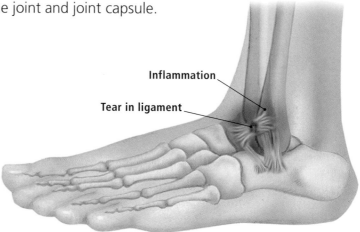

Inflammation

Tear in ligament

Sprain

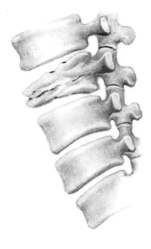

Compression

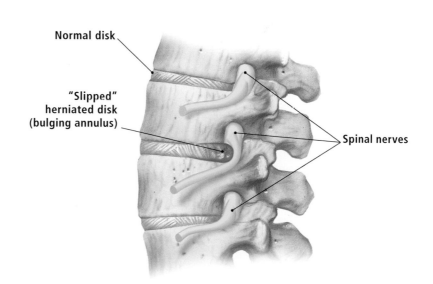

Normal disk

"Slipped" herniated disk (bulging annulus)

Spinal nerves

Herniated disk

Incidental

Incidental stretching is not planned stretching; it is stretching whenever or wherever you see an opportunity. Many people perform these stretches everyday and most probably do them without even thinking about it. Examples of incidental stretches are leaning back on your chair at work, reaching your arms up overhead, or giving yourself a "bear hug." Incidental stretching is informal and therefore can be held for very short periods and is great for breaking up prolonged postures or moving joints that have been held in static positions for long periods.

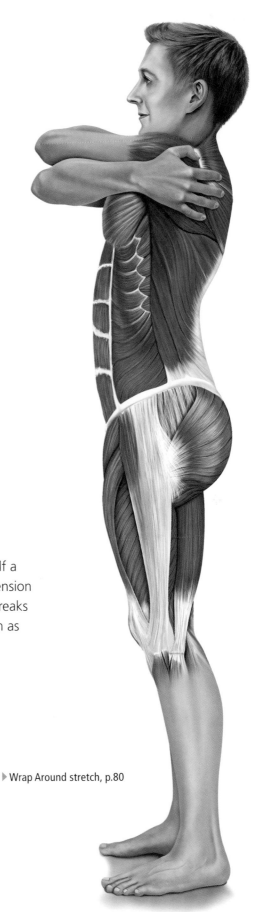

Incidental

Incidental stretching like giving yourself a "bear hug" is a good way to relieve tension in tight muscles. This type of stretch breaks up unnaturally sustained postures such as when we type at a computer.

 Wrap Around stretch, p.80

Safety When You Stretch

Intensity

The intensity of a stretch needs to reach a certain level to be effective while avoiding overstretching that may cause injury or reduce performance. Recent studies have concluded that static stretches of long duration at or near maximum stretch intensity are useful for a range of health-related motion benefits, but are most likely detrimental to force and power production needed in many sports. Sports warm-ups should generally include a submaximal aerobic component followed by large-amplitude dynamic activities and sport-specific activities. Static stretches should form part of the preparation for sports that involve end-of-range static stretch positions, but these should be of short duration and at submaximal intensity.

Pain

As a general rule, stretching should never be painful. Mild discomfort may be expected, but pain is usually a sign that you should immediately stop the stretch. At times, muscles may develop painful contractures, and to relieve these robust—even painful—stretching may be necessary; however, this should only be performed after specific expert instruction.

Processing Pain

The brain controls our perception of pain in a number of different areas. Pain signals are relayed to the brain via the thalamus. The sensation and location of pain is registered in the sensory cortex and emotional responses are governed by the limbic system.

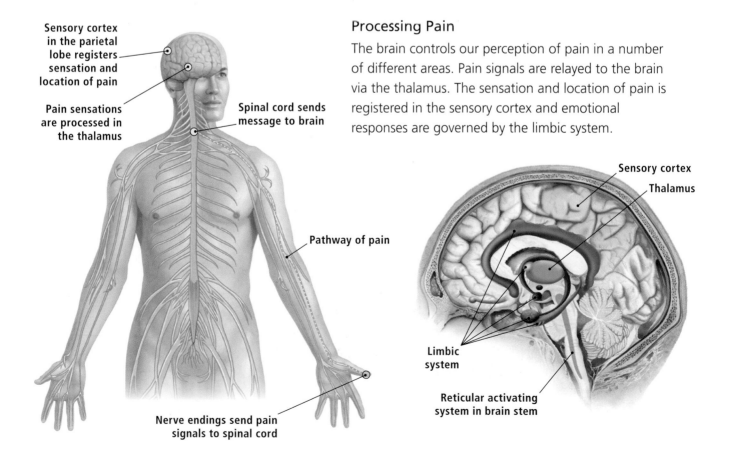

Sensory cortex in the parietal lobe registers sensation and location of pain

Pain sensations are processed in the thalamus

Spinal cord sends message to brain

Pathway of pain

Nerve endings send pain signals to spinal cord

Sensory cortex

Thalamus

Limbic system

Reticular activating system in brain stem

Breathing

Keep breathing while you stretch! Holding your breath can reduce oxygen levels to the muscles and brain, and can be dangerous at worst and disadvantageous at best. Maintaining a relaxed rate of breathing will help reduce muscle tension and increase the benefits of stretching for relaxation or muscle lengthening. During warm-ups, one of the aims of stretching is to increase the rate of oxygen to the working muscles, so maintaining a steady rate of breathing is very important.

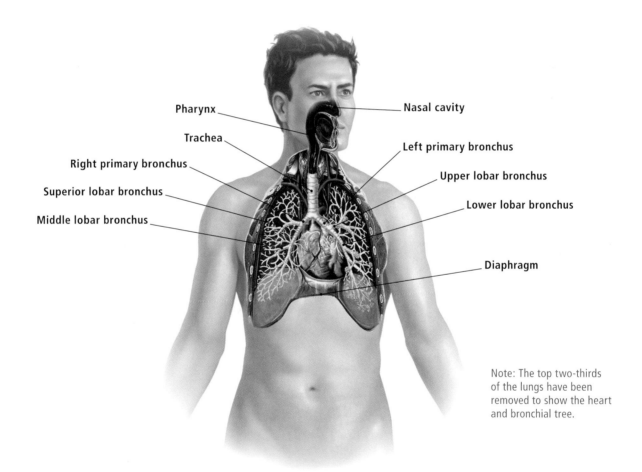

Pharynx

Trachea

Right primary bronchus

Superior lobar bronchus

Middle lobar bronchus

Nasal cavity

Left primary bronchus

Upper lobar bronchus

Lower lobar bronchus

Diaphragm

Note: The top two-thirds of the lungs have been removed to show the heart and bronchial tree.

Respiratory System—Anterior View

Keep breathing while you stretch so that your muscles have a good supply of oxygen as they are extended. A good oxygen supply to the muscles is essential to the warm-up process.

Safety When You Stretch

Dizziness

Certain stretches, particularly neck stretches, may compromise the blood vessels supplying blood and oxygen to the brain. Dizziness while stretching is potentially very dangerous and should be avoided at all costs. If you experience dizziness while stretching, see your health professional as soon as possible.

Balance

Some stretches involve standing in awkward postures. If you have any concerns regarding balance, make sure that you are within reach of something to grab hold of while performing standing stretches. Remember that, for static stretches, it is vital that the targeted muscle is held in a constant stretch position. If you are hopping around trying to maintain balance, the stretch will no longer be of any benefit!

Balance

Specialized organs in the inner ear known as the semicircular canals and the otolith organs contain tiny hairs that are sensitive to the body's position in space. Changes in position excite the hairs, which send nerve signals via the vestibular nerve to the brain. The brain uses this information to help balance the body.

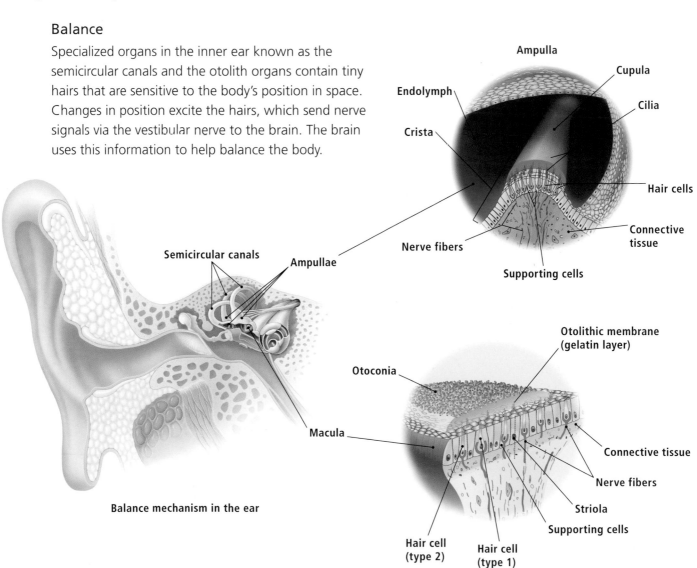

Balance mechanism in the ear

Frequency

There is good evidence that stretching too frequently following an injury can lead to higher rates of injury on returning to sport. One study of Greek soccer players found that those who stretched more than twice a day had a higher rate of injury recurrence. It seems that too frequent stretching produces a mistaken sense of recovery following acute muscle strain and encourages a return to sport when muscles have recovered their length but not their strength.

Stretching after injury

Stretching after injury should always be undertaken under the supervision of a health professional or trained exercise consultant. Too frequent stretching may increase the risk of injury recurrence.

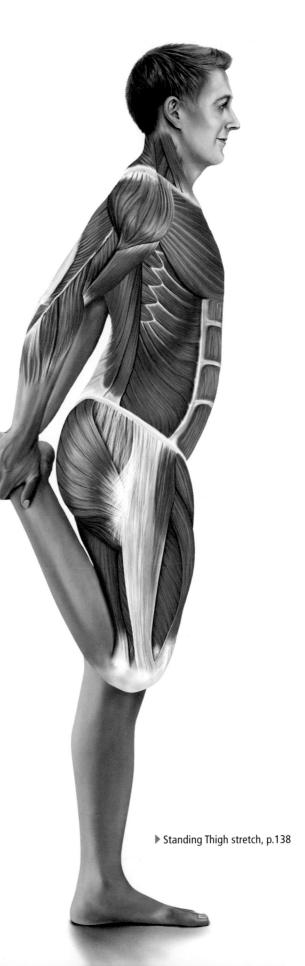

▶ Standing Thigh stretch, p.138

How to Stretch Properly

There is no definitive "proper way to stretch": even the experts can't agree on that. However, what can be said with confidence is that what is best for you depends on your circumstances and your intentions. These are some of the most commonly recommended stretching routines for particular goals.

For warm-up

If you want to incorporate stretches into a sport or exercise warm-up, then they need to be of the dynamic kind. As mentioned earlier, static stretching does nothing to make muscles warmer. Dynamic stretching involves moving muscles and joints through movements that are similar to the activity about to be performed. Move at a slower intensity than the upcoming activity to warm up the muscles and to prepare the mind for the upcoming activity. Simply walking and slow jogging before running would be a perfect form of warm-up; but, for more complex sports, a routine of dynamic stretching and slow-paced skills performance would be more appropriate.

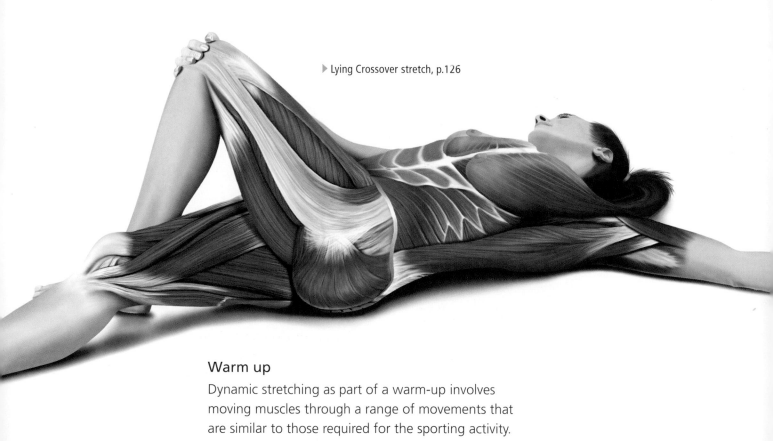

▶ Lying Crossover stretch, p.126

Warm up
Dynamic stretching as part of a warm-up involves moving muscles through a range of movements that are similar to those required for the sporting activity.

For flexibility and improved range of motion

If you are determined to improve your flexibility or need to be more flexible to perform certain activities, then sustained static stretching is best. To do this, you need to stretch the joint to its end-of-range position and hold it there for a prescribed length of time. There is no "golden rule" as to how long to hold a stretch; but, generally accepted practice states that, the more often you stretch and the longer you hold a stretch for, the faster your flexibility is likely to improve. Some research suggests that the improvements from static stretching are not actually a change in muscle length, but result from improved stretch pain tolerance that enables a greater strain to be placed on the muscle or joint. Whether the change is related to altered muscle length or stretch tolerance, the desired increase in joint range of motion can be achieved with a bit of dedication. However, be sure that this is appropriate for your situation, as hypermobile joints can be more detrimental than beneficial in some situations and overstretching could possibly cause injury.

Flexibility and Movement

Stretching can improve flexibility and range of movement. You need to make sure that increased range of joint movement won't be detrimental for your desired activity.

▶ Rotating Wrist stretch, p.94

How to Stretch Properly

To improve posture

Many people stretch in the hope of improving their posture. Static stretches, dynamic stretches, or mobilizations are all likely to do an equally good job. Postural stretches are usually targeted around the neck, upper back, shoulders, and hips. The aim of stretching to improve posture is to optimize the kyphosis and lordosis of the spine, minimize a "poked head" position, and reverse the "rolled" shoulder and rounded back slouch. An improved posture is beneficial for breathing and oxygen delivery, reducing back and neck pain, and general well-being and confidence.

For relief of static and prolonged postures

Many people in their work or sport find themselves in the same position or posture for hours. Prolonged static postures can lead to injuries or general aches and pains. Stretching and mobilization techniques can be very effective in avoiding or relieving these problems. The main aim is to reverse the prolonged posture at regular intervals. For example, an office worker who sits at a desk all day should stand up every hour or so and arch backwards, roll the shoulders back, reach for the sky, and tilt his or her head side to side. There are no rules as to how many times or how long to hold these stretches other than to say that the longer you spend between changing positions, the longer you should spend on your stretching routine.

Static Posture Relief

Stretching has benefits for relieving the static or prolonged postures that we adopt during our regular daily activities. The child pose helps to mobilize stiff joints and relieve aches.

▶ Child Pose, p.114

For relaxation

We all stretch at times simply because it feels good. There are no rules as to how to stretch for relaxation—do what feels right and makes you feel good! Remember to breathe through your stretching, as slow rhythmic breathing is also beneficial for relaxation, while holding your breath can be dangerous.

Spinal curvature

The spinal column is normally curved, but abnormal or excessive curvature (as in excess kyphosis or scoliosis) may be caused by disease, injury, or congenital abnormality.

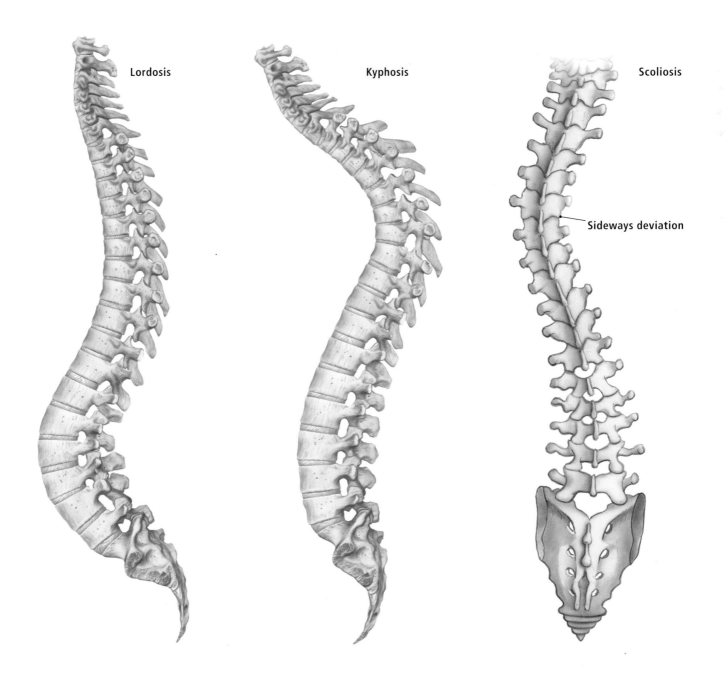

Lordosis

Kyphosis

Scoliosis

Sideways deviation

Stretching for Special Groups

For injury rehab/prehab

Stretching plays a vital role in the field of injury rehabilitation. Following injury, joints may become tighter and lose range of motion, muscles can become tighter or shorter, and trigger points can develop. This can lead to muscle imbalances and ongoing injury or performance issues. In this situation, your health professional can identify these muscle imbalances and instruct you on appropriate and specific stretches. "Prehab" relates to identifying potential areas of concern and addressing them before they become injuries. Muscle imbalances or suboptimal ranges of motion may be identified and addressed with specific stretches. Special exercises and joint mobilization are key to recovery from knee replacement surgery.

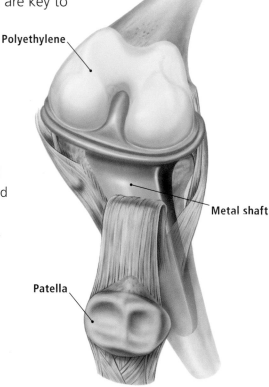

Polyethylene

Metal shaft

Patella

Knee Replacement

The knee joint is particularly vulnerable to stress injuries, and reconstruction or replacement may be necessary if the ligaments have been badly torn or severed. In knee replacement, the damaged joint is repaired by inserting metal shafts into the tibia and femur. A strong polyethylene coating covers the end of the shafts (in place of cartilage) and the knee ligaments are reattached to hold the joint together.

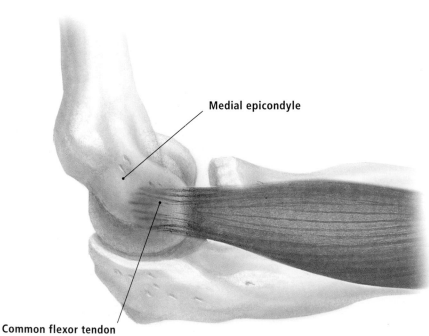

Medial epicondyle

Common flexor tendon

Overuse Injury

Like tennis elbow, golfer's elbow is an injury caused by overuse of particular muscles in the forearm. In golfer's elbow, pain is felt at the inner side of the elbow, where the common flexor tendon inserts into the medial epicondyle.

For children

Many children suffer from growing pains, particularly during periods of rapid growth. Growing pains arise when muscles or other soft tissue structures are forced to adapt to changes in bone length. A regular gentle stretching routine can be very helpful in relieving the pain associated with these growth spurts.

For the elderly

As we age, our soft tissue becomes less elastic and joints typically become stiffer. Identifying the problem areas and adopting safe stretching routines can help to maintain optimal range of motion and reduce the pain associated with stiff joints.

Osteoarthritis

The cause of osteoarthritis is disintegration of the cartilage that covers the ends of the bones. This will occur as part of the normal ageing process and frequently affects the knees, hips, joints of the big toes, and lower sections of the spine. When symptoms are severe, movement can be restricted and part of the affected bone may wear away.

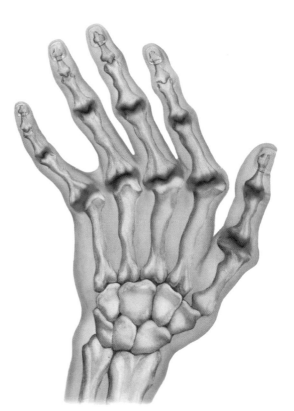

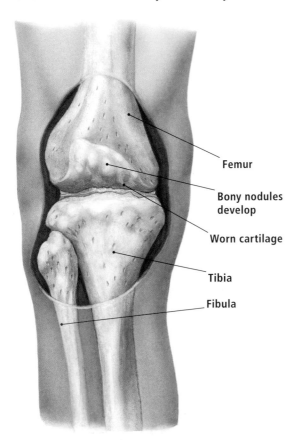

Femur

Bony nodules develop

Worn cartilage

Tibia

Fibula

Rheumatoid Arthritis

One of the most common areas of the body to suffer rheumatoid arthritis is the hands. The joints become stiff, painful, inflamed, and swollen, making even the most simple of tasks—such as picking up an object—difficult or impossible to do.

Stretching for Special Groups

For athletes

Dynamic stretching can be an important part of a pre-exercise warm-up. Some athletes need to be able to achieve ranges of motion that are beyond that which is necessary for everyday living, and vigorous stretching is often necessary to achieve this. Stretching following exercise can help relieve the feeling of stiffness often perceived following workout or competition. There are some very important points to remember though: stretching will not reduce injury or delayed onset muscle stiffness (DOMS), it won't help post-exercise recovery, and it may inhibit performance. There is good evidence that static stretching reduces sprint performance and that stretching before or during strength training reduces strength gains.

Ankle Fracture

The ankle bones are prone to fracture, because the ankle is a weight-bearing joint and is easily put under stress. The ankle fracture shown here is called a Pott's fracture.

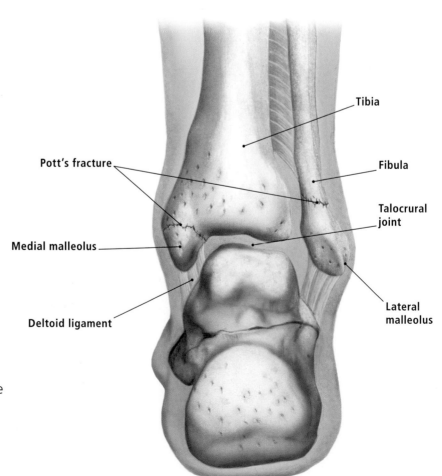

Tibia

Fibula

Talocrural joint

Lateral malleolus

Pott's fracture

Medial malleolus

Deltoid ligament

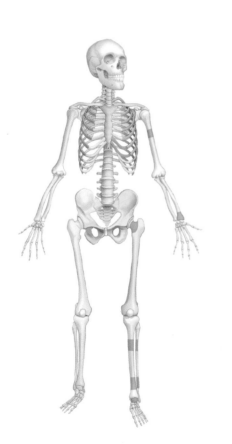

Common Fracture Sites

Common fracture sites include the arm, wrist, hip, leg, and ankle.

For the immobile

People who are wheelchair bound, confined to bed, or lack active muscle control of a body area will benefit from regular stretching. Having a joint or joints that do not change position leads to muscle contractures. While short periods of stretching are not enough to reverse these contractures, they still provide some relief to chronically stiff joints. Devices such as night splints can be used for longer stretch periods if desired. People who perform regular stretching in these situations report feeling much better, but the stretching needs to be regular and ongoing.

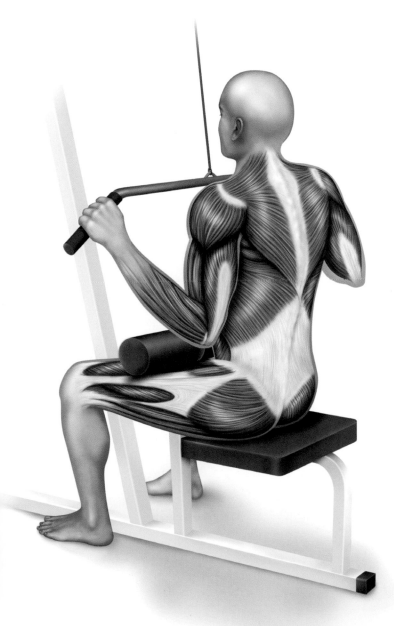

T1

Spinal cord

Spine

L1

L2

Paralysis—Paraplegia

Paraplegia is the result of injury or disease to the spinal cord between the T1 (thoracic) and L2 (lumbar) segments. It spares the arms, but depending on the nerves damaged may involve the legs, pelvic organs, and trunk.

Basic Stretches

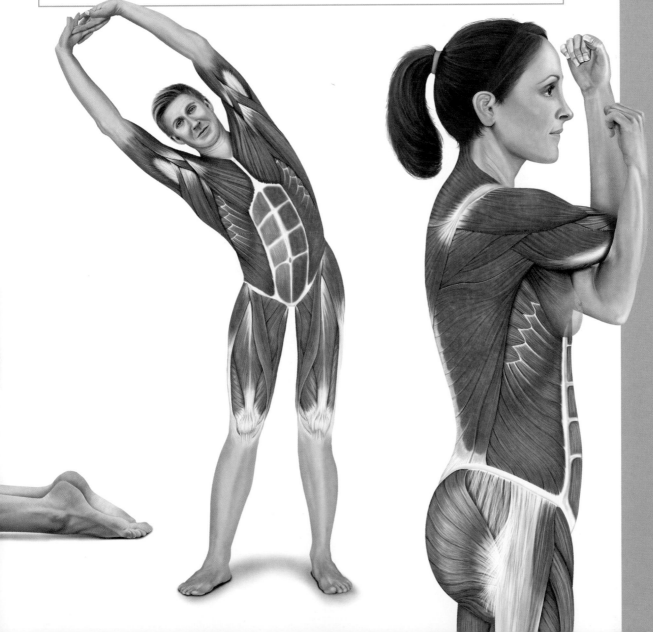

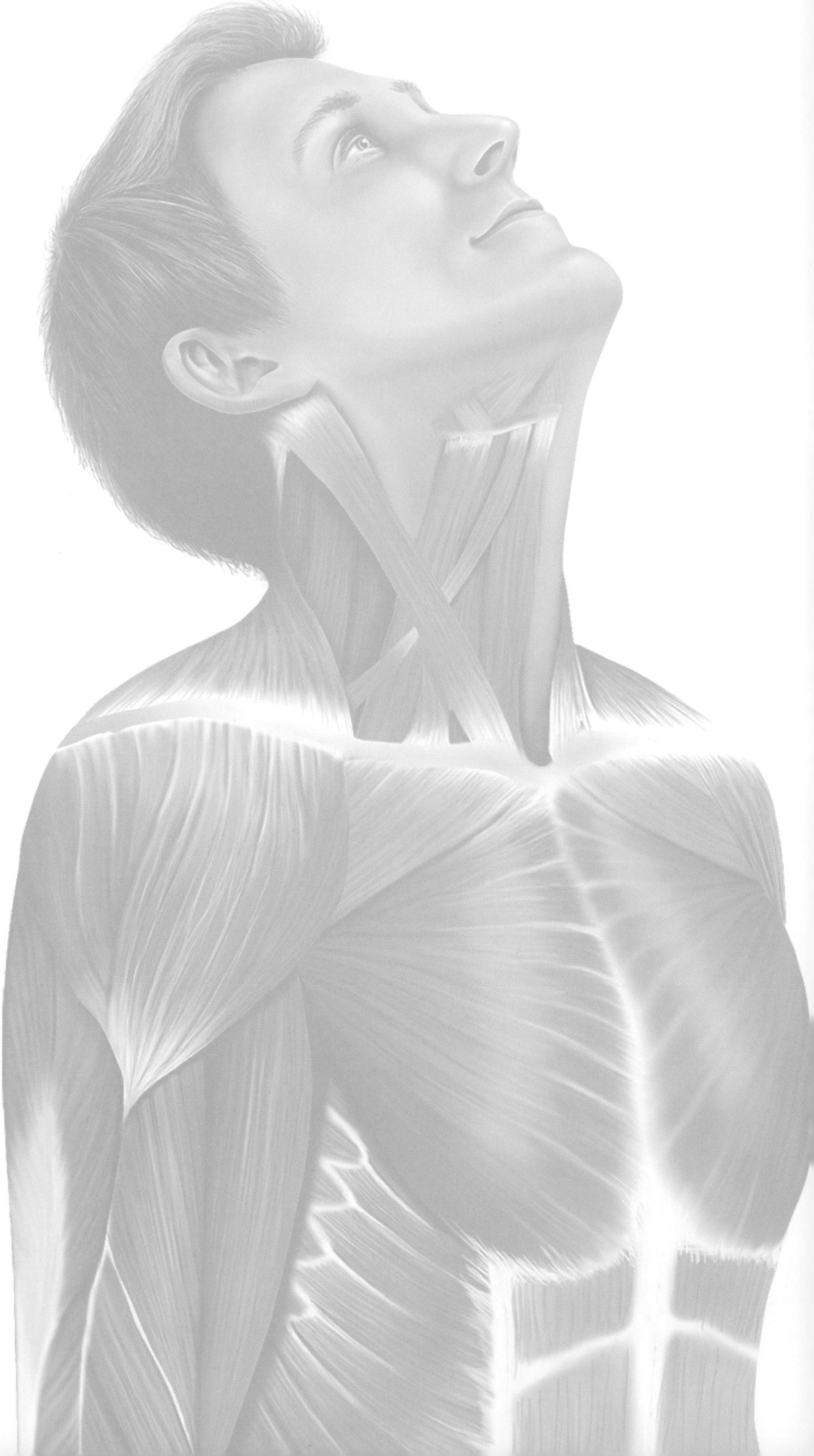

Neck Stretches

The muscles of the neck work together to move the head in every direction. They also assist in the control of functions such as swallowing and facial expression. People often experience neck stiffness after carrying heavy bags, sleeping awkwardly, or from exercising with incorrect technique. Extended computer use or sitting for prolonged periods with poor posture may increase the risk of shoulder and neck disorders. This may lead to reduced strength and function of the neck or increase frequency of headaches. The right stretching program may help reduce tightness and, when combined with exercise, may help to correct poor posture. Neck stretches should be performed carefully and slowly, especially if you have any previous injuries or any degeneration of the cervical spine.

Muscles and Joints of the Neck

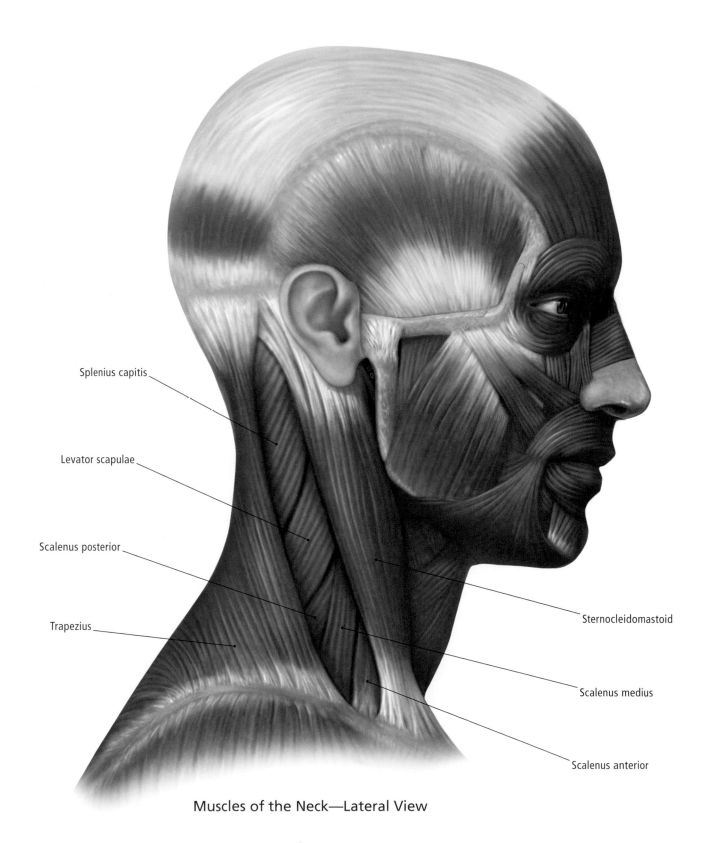

Splenius capitis

Levator scapulae

Scalenus posterior

Trapezius

Sternocleidomastoid

Scalenus medius

Scalenus anterior

Muscles of the Neck—Lateral View

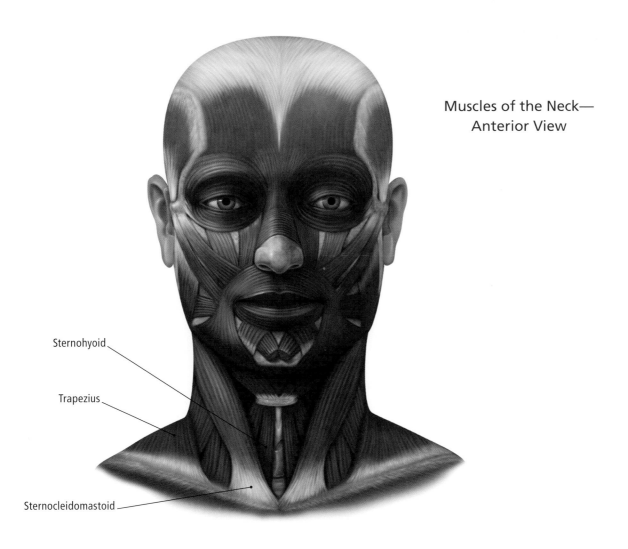

Muscles of the Neck—
Anterior View

Sternohyoid

Trapezius

Sternocleidomastoid

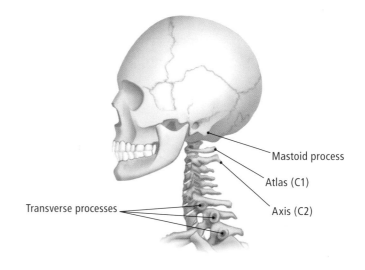

Mastoid process

Atlas (C1)

Axis (C2)

Transverse processes

Bones of the Neck—Lateral View

Rotating and Lateral Neck Stretches

We rely on a high level of neck mobility to orient our senses while playing sport and during everyday life, so the muscles that control lateral movements and rotation of the head and neck are being used all the time. Unbalanced exercise, maintaining poor postures for prolonged periods, jolts caused by collision, or even stress, can cause stiffness and tightness in the neck due to the extra strain put on these muscles. These stretches can be performed anywhere without any equipment, and may help relieve stiffness in the neck and restore range of motion. There are a few instances in which neck stiffness can indicate a serious medical condition, so check with a health professional if this persists.

do it right

Keep your hands behind your back to help keep your shoulders still while doing these stretches.

how to

For the rotating neck stretch, begin with an upright posture in either a standing or sitting position while looking straight ahead. Keep your chin at the same level, and slowly turn your head to one side as though you are going to look over one shoulder.

For the lateral neck stretch, begin by standing or sitting upright while looking straight ahead. Keep your shoulders down and back, and slowly drop your ear toward your shoulder. Continue to face forward.

variations

ONE A If you are free from any previous neck injury, you can accentuate the stretch by varying your head position. Keep your chin at the same level, slowly turn your head to one side as though you are going to look over one shoulder, and slowly tilt your head slightly backward.

ONE B If you have neck stiffness higher up on your neck, closer to the base of the skull behind your ear, you can focus the lateral stretch a little more by dropping your ear toward your shoulder. Then slowly turn your head slightly to look down toward the floor.

warning

Slowly move in and out of these stretches and do not roll your head from one position to the other. If you experience any severe pain, you should stop the stretch and consult a health professional.

muscles being stretched

❶ Sternocleidomastoid

❷ Scalenus posterior

❸ Scalenus medius

❹ Scalenus anterior

❺ Levator scapulae

❻ Trapezius

❼ Splenius capitis

❽ Longissimus capitis

❾ Semispinalis capitis (7, 8, and 9 all under Trapezius)

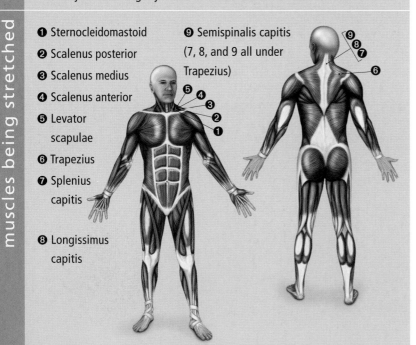

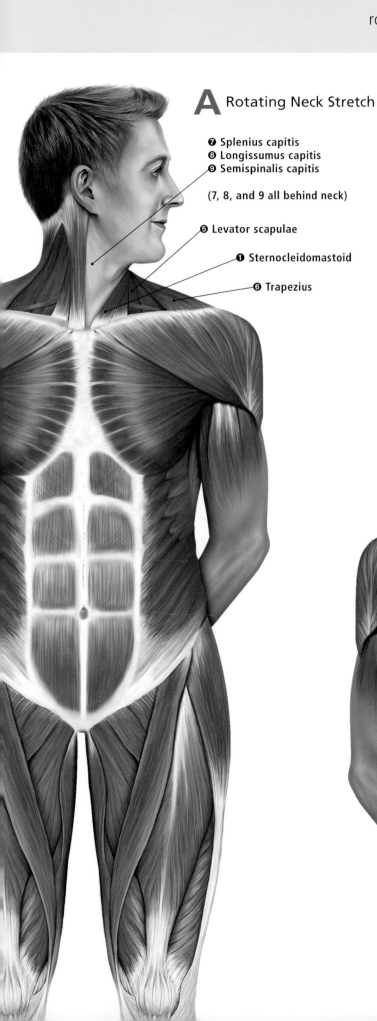

A Rotating Neck Stretch

❼ Splenius capitis
❽ Longissumus capitis
❾ Semispinalis capitis

(7, 8, and 9 all behind neck)

❺ Levator scapulae

❶ Sternocleidomastoid

❻ Trapezius

B Lateral Neck Stretch

❶ Sternocleidomastoid

❺ Levator scapulae

❹ Scalenus anterior

❸ Scalenus medius

❷ Scalenus posterior

❻ Trapezius

Forward Neck Flexion and Neck Protraction

Tightness and stiffness at the back of the neck is common among people who spend a lot of time at a desk or in front of a computer. With your neck in a forward position and shoulders hunched forward, the neck extensor muscles at the back of the neck have to work harder to maintain a forward gaze. This can lead to shoulder problems and injury, difficulty sleeping, and headaches. Having strong, well functioning neck extensors is also very important for those involved in contact sports, particularly football, rugby, and wrestling, where they act to protect the spine from damage. Stretching and strengthening these muscles may restore range of motion where it has been lost, improve posture, and protect the cervical spine from injury.

warning

In most people, these muscles are weak and require strengthening, so ensure you don't overstretch this area. See a health professional if you have ongoing stiffness or pain.

how to

For the forward neck flexion stretch, stand or sit with your arms by your sides. Start fully upright—imagine a piece of string is pulling the crown of your head toward the ceiling. Then drop your chin toward your chest by curling the neck forward, starting with the vertebrae at the very top, followed by the ones underneath. To do this, first pull your chin back, like you have a double chin, before flexing forward.

For the neck protraction, begin with an upright posture in either a standing or sitting position looking straight ahead. While keeping your chin at the same level, slowly thrust your head forward without tilting the head down.

variations

ONE **A** If you have tightness in one side more than the other, or you need to focus the forward neck flexion on one side at a time, then once you have dropped your chin to your chest, slowly tilt your head, taking your ear toward your shoulder.

TWO **A** You can use one hand to gently assist with the neck flexion stretch or with its variation above. Use only gentle pressure and don't hold the stretch for prolonged periods.

muscles being stretched

❶ Levator scapulae

❷ Trapezius

❸ Splenius capitis

❹ Splenius cervicis

❺ Spinalis capitis

❻ Spinalis cervicis

❼ Semispinalis capitis

❽ Semispinalis cervicis

❾ Longissimus capitis

❿ Longissimus cervicis (3 to 10 all under Trapezius)

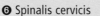

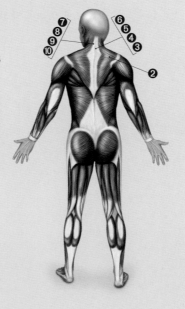

do it right

Move slowly with these stretches and ensure you continue to breathe normally throughout the stretch.

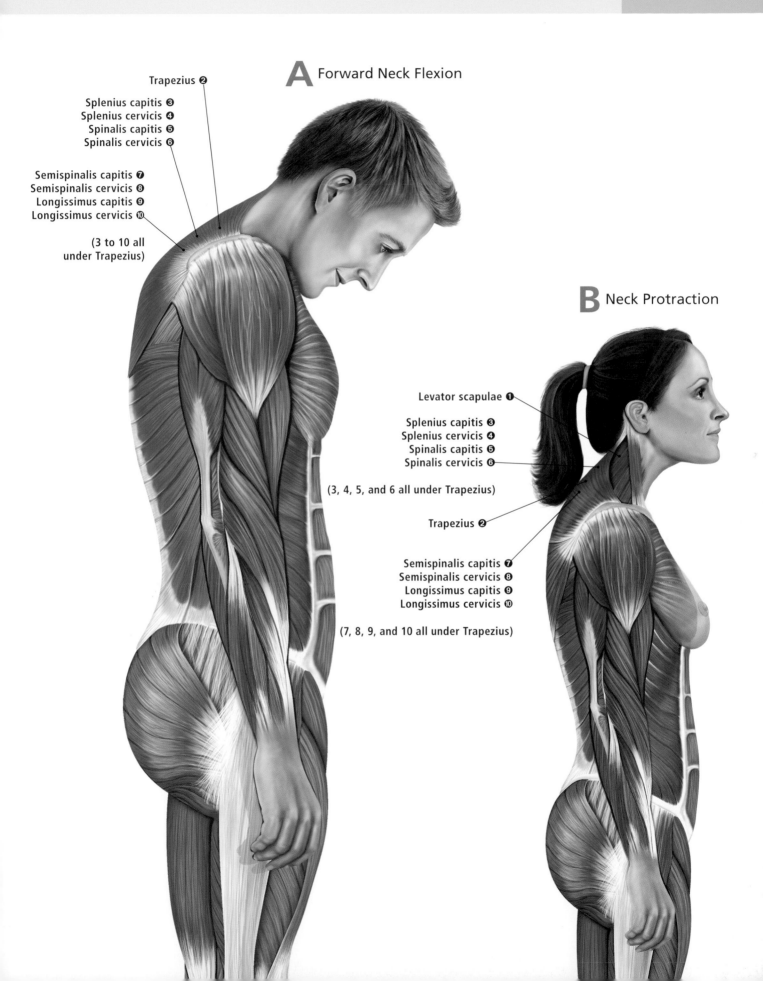

A Forward Neck Flexion

Trapezius ❷

Splenius capitis ❸
Splenius cervicis ❹
Spinalis capitis ❺
Spinalis cervicis ❻

Semispinalis capitis ❼
Semispinalis cervicis ❽
Longissimus capitis ❾
Longissimus cervicis ❿

(3 to 10 all under Trapezius)

B Neck Protraction

Levator scapulae ❶

Splenius capitis ❸
Splenius cervicis ❹
Spinalis capitis ❺
Spinalis cervicis ❻

(3, 4, 5, and 6 all under Trapezius)

Trapezius ❷

Semispinalis capitis ❼
Semispinalis cervicis ❽
Longissimus capitis ❾
Longissimus cervicis ❿

(7, 8, 9, and 10 all under Trapezius)

Neck Extension

This stretch targets the neck flexor muscles that pull the chin and head toward your chest. These muscles also work with other muscles to coordinate movements such as rotating the head from side to side. With prolonged sitting, the head and shoulders can drift forward, causing these muscles to become tight. This can perpetuate postures that cause neck pain. This stretch is excellent for improving the range of motion in the neck flexors, and can be performed at any time by most people. As with all neck stretches, this should be performed slowly and gently. If you have had a previous neck injury, it is important to consult a health professional before trying this stretch.

warning

Do not roll your head from side to side while your neck is extended. Stretch each side separately if required.

how to

Maintain an upright posture, tilt your head backward, and look up toward the ceiling. In order to avoid bunching your neck and shoulders up at the back, picture yourself thrusting your chin toward the ceiling. This stretch can be performed while standing or seated.

variation

ONE

A From time to time, you may feel tighter or stiffer on one side more than the other. This may be due to an activity that is dominated by one side, such as golf or tennis, or by holding an unbalanced posture for a prolonged period, such as when you are sleeping. Whatever the reason, if this is the case, you can vary the stretch to accentuate one side over the other. Tilt your head backward and look up toward the ceiling. Focus the stretch on one side by tilting your head slightly away from that side. For example, thrust the right side of your chin toward the ceiling to feel the stretch more on that side.

muscles being stretched

❶ Sternocleidomastoid

❷ Omohyoid

❸ Sternohyoid

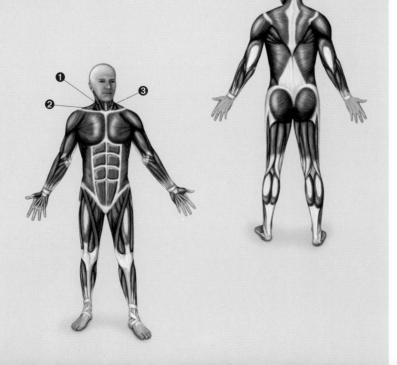

A Neck Extension

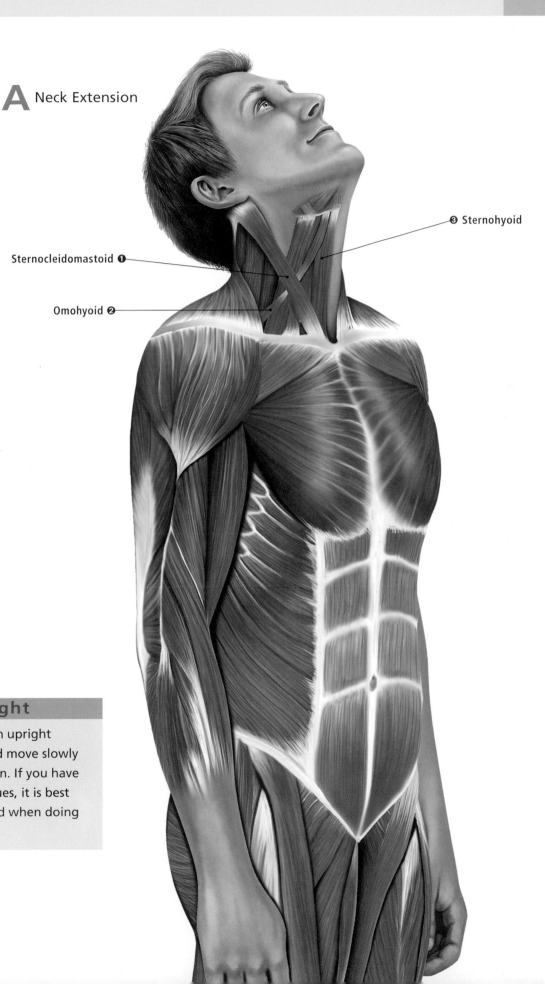

❸ Sternohyoid

Sternocleidomastoid ❶

Omohyoid ❷

do it right

Maintain an upright posture and move slowly into position. If you have balance issues, it is best to be seated when doing this stretch.

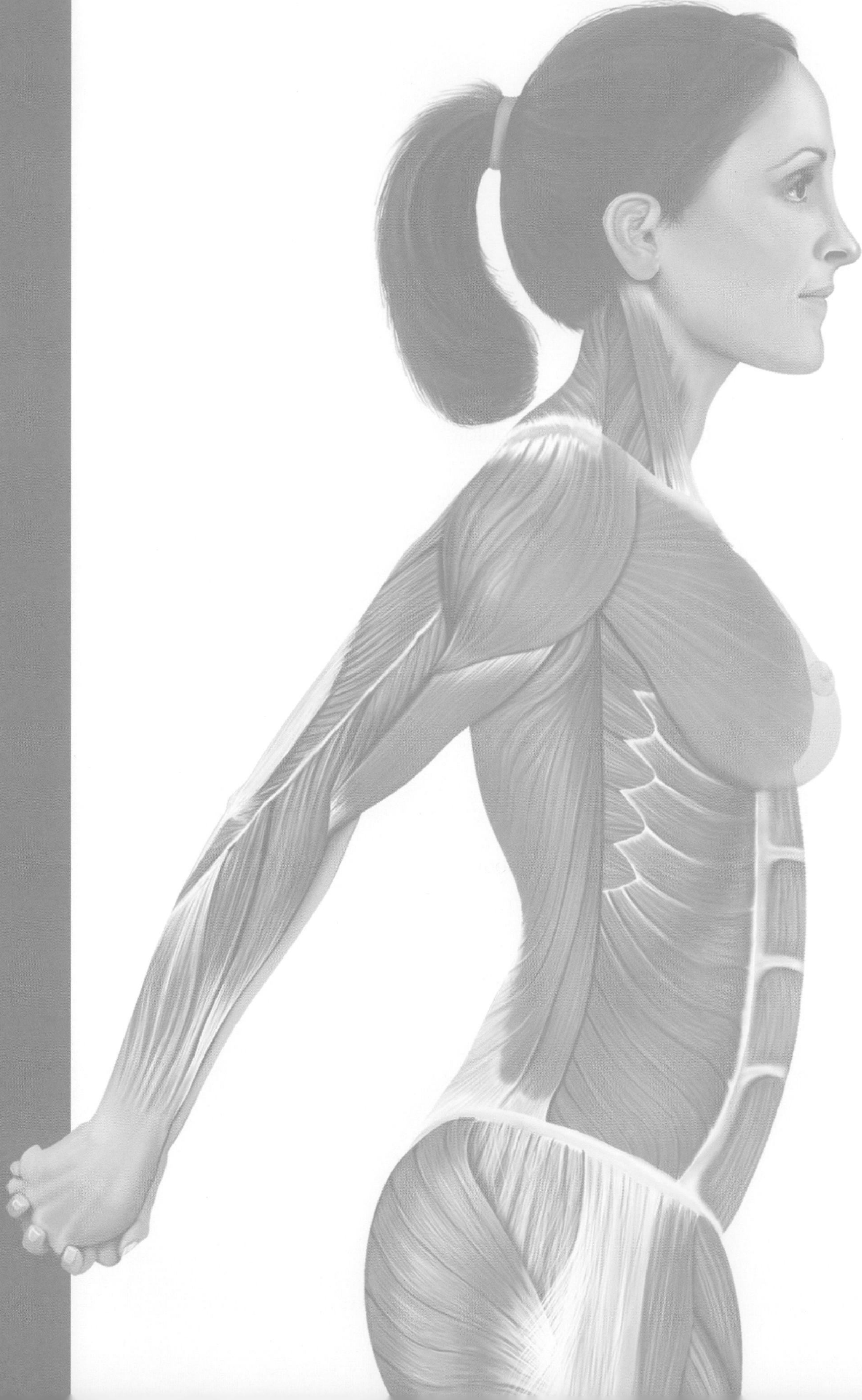

Shoulder Stretches

The musculature of the shoulder is made up of large muscles that produce movement, as well as smaller muscles that control and stabilize the joints. As the shoulder forms the link between the torso and the arms, these muscles are involved in many day-to-day and sporting activities. Poor posture or an unbalanced exercise program may lead to excessive tightening, particularly in the anterior shoulder region. This can lead to longer-term problems that may impair balance, the ability to carry out everyday activities, and even affect breathing. The right stretches performed correctly can help to alleviate tightness and restore function. Shoulder stretches are particularly important for sports that involve high velocity movements like throwing, in order to avoid internal impingement.

Joints and Muscles of the Shoulder

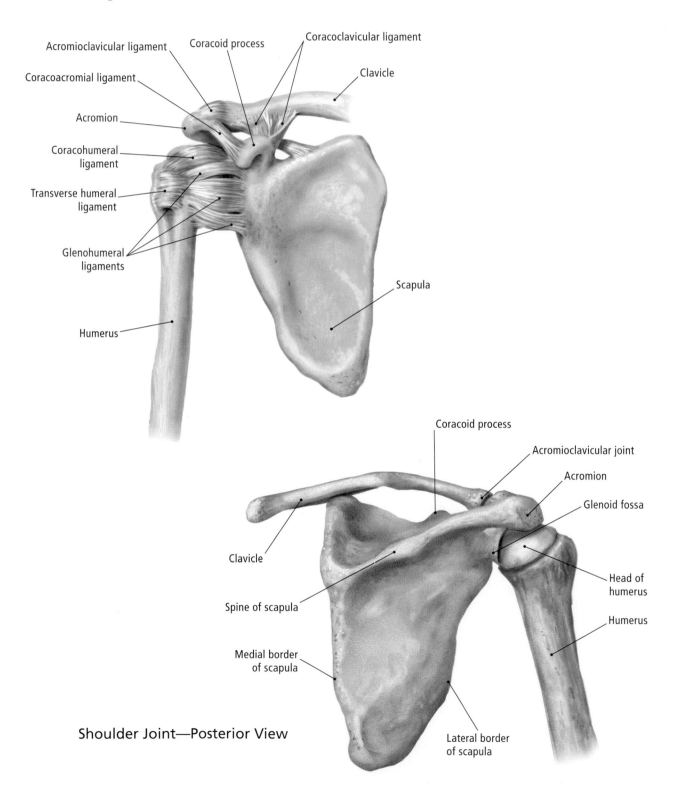

Ligaments of the Shoulder—Anterior View

Acromioclavicular ligament

Coracoid process

Coracoclavicular ligament

Clavicle

Coracoacromial ligament

Acromion

Coracohumeral ligament

Transverse humeral ligament

Glenohumeral ligaments

Scapula

Humerus

Coracoid process

Acromioclavicular joint

Acromion

Glenoid fossa

Clavicle

Head of humerus

Spine of scapula

Humerus

Medial border of scapula

Lateral border of scapula

Shoulder Joint—Posterior View

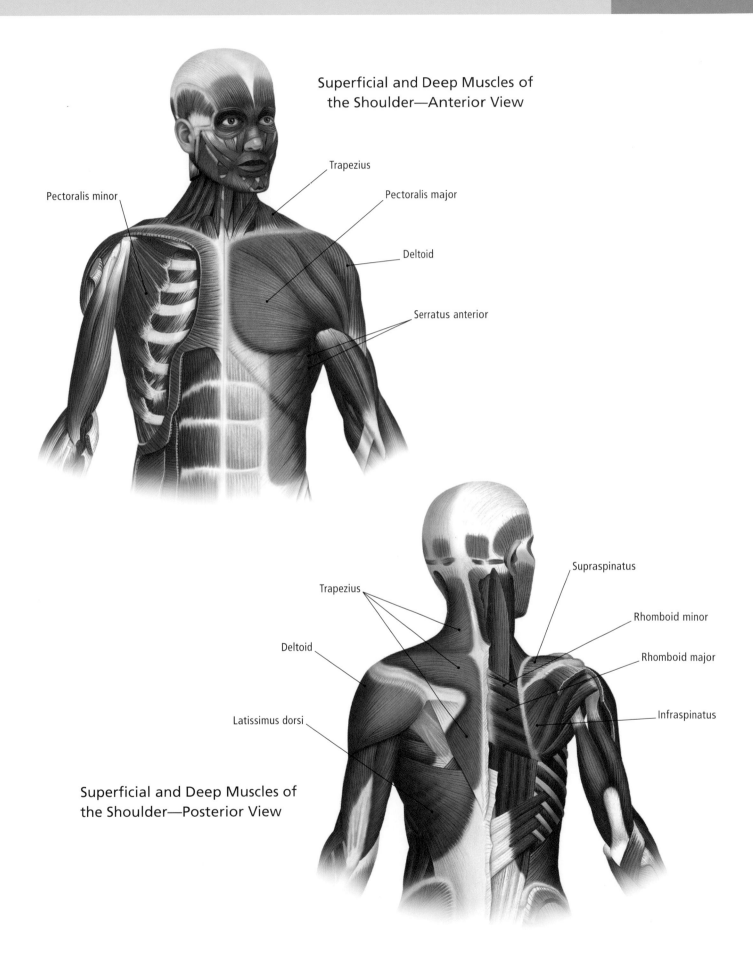

Superficial and Deep Muscles of the Shoulder—Anterior View

Pectoralis minor

Trapezius

Pectoralis major

Deltoid

Serratus anterior

Trapezius

Supraspinatus

Deltoid

Rhomboid minor

Rhomboid major

Infraspinatus

Latissimus dorsi

Superficial and Deep Muscles of the Shoulder—Posterior View

Cross-body Shoulder Stretch

This stretch, suitable for most people, can be done in a standing or sitting position and requires no equipment. It targets the muscles at the rear of the shoulder and in the upper back between the shoulder blades, predominately the posterior deltoid and rhomboids. Often these muscles are weak in people with a poor, forward-slumped posture and require strengthening to restore balance. This stretch may relieve tightness after these muscles have been targeted in a workout. This stretch may also be beneficial after activities that involve a lot of pulling, lifting, or carrying such as kayaking, martial arts, or even carrying heavy shopping bags.

warning

Be careful not to force the stretch by pulling too vigorously with your other arm.

how to

Raise the arm you wish to stretch up to shoulder height and bring it across your body. Keep the upper surface of your arm parallel to the ground and your fingers pointed straight up. Bring your other arm up and use it to gently pull the stretching arm across and toward your body. Avoid rotating your torso or slumping your head forward.

variations

ONE

A To focus the stretch more on the rhomboids, stretch both sides at the same time by holding both arms outstretched. Rotate your hands so that your thumbs are pointing toward the floor, then cross your hands over so that your palms are facing each other. Reach your arms forward to stretch or grasp a doorframe or fixed pole between your hands and lean back to feel the stretch.

TWO

A If you find it uncomfortable to have both arms across your body, or if you feel like your chest area is restricting you and stopping you from stretching the target area, try keeping your arm straight and use a wall or other solid object to hold your arm in position while you stretch, leaving your other arm by your side.

muscles being stretched

❶ Trapezius

❷ Infraspinatus

❸ Teres minor

❹ Triceps brachii

❺ Latissimus dorsi

❻ Rhomboids (under Trapezius)

❼ Posterior deltoid

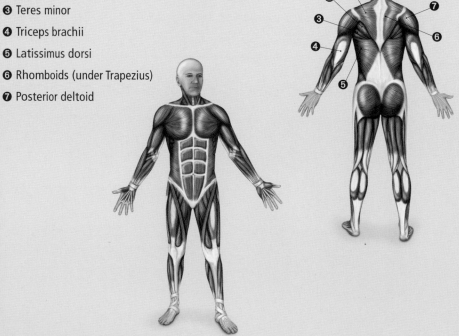

A Cross-body Shoulder Stretch

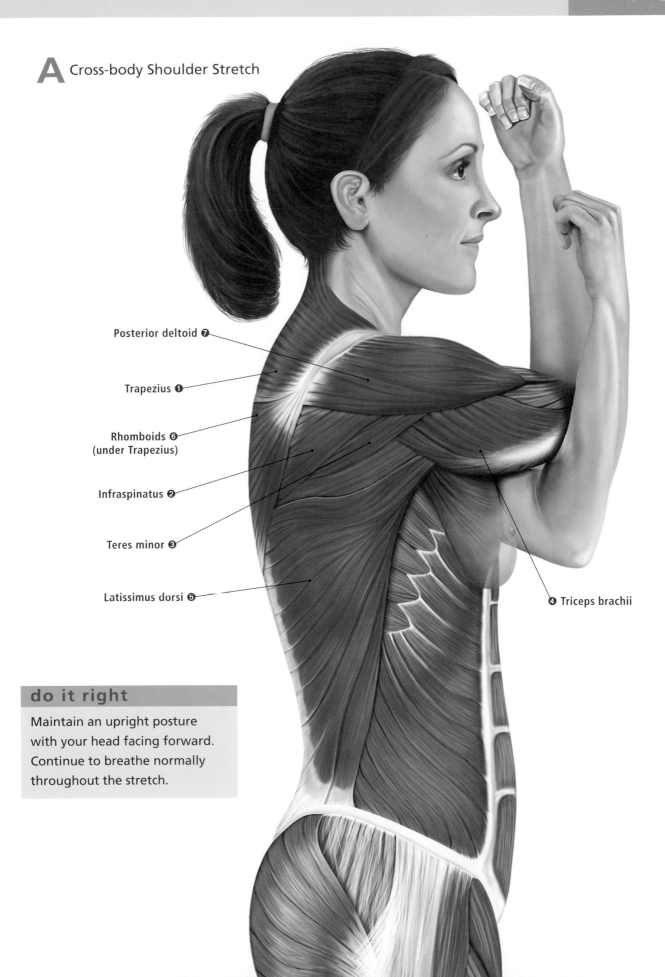

Posterior deltoid ❼

Trapezius ❶

Rhomboids ❻
(under Trapezius)

Infraspinatus ❷

Teres minor ❸

Latissimus dorsi ❺

❹ Triceps brachii

do it right

Maintain an upright posture
with your head facing forward.
Continue to breathe normally
throughout the stretch.

Rotator Cuff Stretch

The shoulder joint is highly mobile, allowing a large range of movement and dexterity. In order to provide stability, a group of muscles, collectively known as the rotator cuff, work together from different angles to keep the head of the humerus securely in place. This stretch focuses on two of these muscles—infraspinatus and teres minor— that externally rotate the humerus. These muscles are particularly important to athletes that do a lot of throwing, as they provide a braking force to slow the arms down after the throw. This stretch can be done with no equipment and should be performed slowly and gently.

how to

Place the back of your hand against the small of your back, keeping your elbow close to your body. From this position, standing with an upright posture or slightly arching your shoulders backward may be enough to feel this stretch. If not, bring your other arm across in front of your body and gently pull your elbow forward.

variations

ONE

A Place your hand on your hip so that your elbow is pointing out to the side and your arm is in line with your body. Reach your other arm across in front of your body and gently pull your elbow forward. This will change the direction of stretch on the target muscles and may involve teres minor to a greater degree.

TWO

A Place the back of your hand against the small of your back, keeping your elbow close to your body. Leaving your other arm by your side and looking straight ahead, slowly reach your thumb up your spine to provide the stretch. This variation may be easier if you find it difficult to reach across your body with the other arm.

muscles being stretched

❶ Teres minor

❷ Infraspinatus

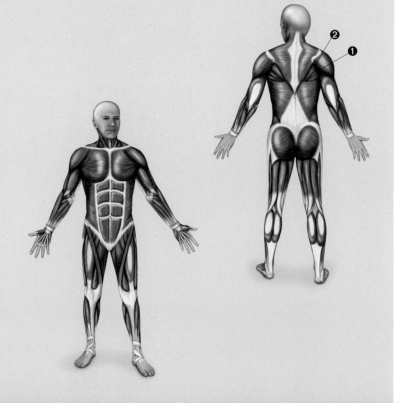

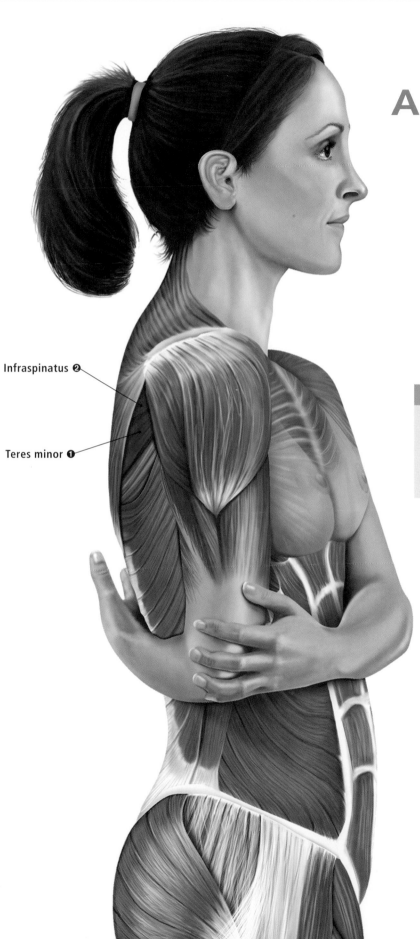

A Rotator Cuff Stretch

Infraspinatus ❷

Teres minor ❶

do it right
Maintain an upright posture, looking straight ahead with your shoulders back.

Assisted Infraspinatus and Subscapularis Stretches

These stretches target the muscles that stabilize the shoulder joint, known as the rotator cuff. The assisted infraspinatus stretch focuses on the muscles that externally rotate the arm—the movement generated during a tennis backhand. The assisted subscapularis stretch focuses on the muscles that internally rotate the arm—the movement generated during a tennis forehand or an arm wrestle. These muscles, however, are active during every movement at the shoulder joint. As such, the rotator cuff is vulnerable to overuse, muscle imbalance, and injury, and is a common cause of disability. A program of stretching and strengthening can help to protect the shoulder and maintain function.

warning

Do not perform these stretches without seeking the advice of a health professional if you have had a previous shoulder injury.

how to

For the assisted infraspinatus stretch, hold the end of a pole in one hand so that the long end is running toward your little finger, the short end toward your index finger. Hold the pole by your side so that it runs up behind your arm and shoulder. Use your other hand to pull the top of the pole forward until you feel the stretch behind your shoulder.

For the subscapularis stretch, grasp the end of a pole in one hand so that the long end is facing up. Bring you arm up so that your elbow is level with your shoulder and the pole is running down the middle of the back of your arm toward the floor. Reach across your body with your other arm and gently pull the pole forward.

variation ONE

A You can perfom the subscapularis stretch without the use of a pole by using another solid object such as a wall or doorway. While standing upright, bend your right elbow to ninety degrees, keeping your elbow by your side. Place your hand against the doorway and gently externally rotate your arm by turning your body to the left.

muscles being stretched

❶ Long head of biceps brachii

❷ Subscapularis (on the front surface of the scapula)

❸ Pectoralis major

❹ Anterior deltoid

❺ Infraspinatus

❻ Teres minor

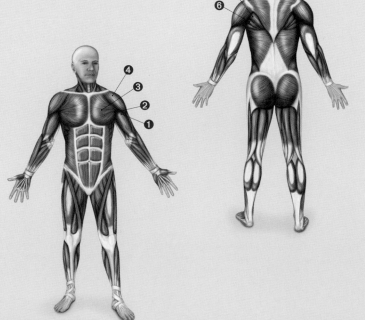

do it right

Apply gentle pressure with the assisting arm to avoid overstretching or injury.

A Assisted Infraspinatus Stretch

B Assisted Subscapularis Stretch

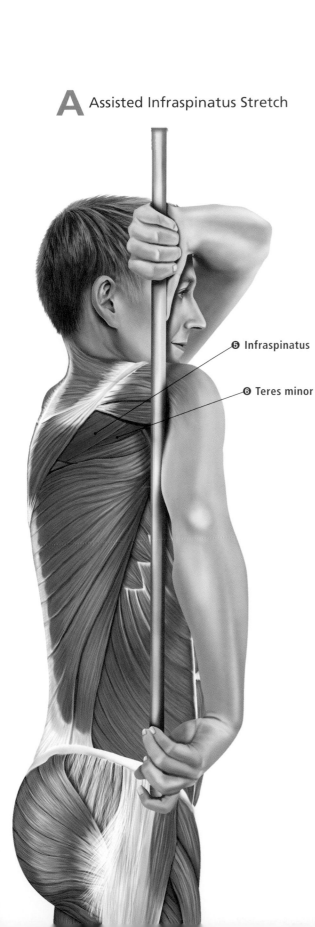

❺ Infraspinatus

❻ Teres minor

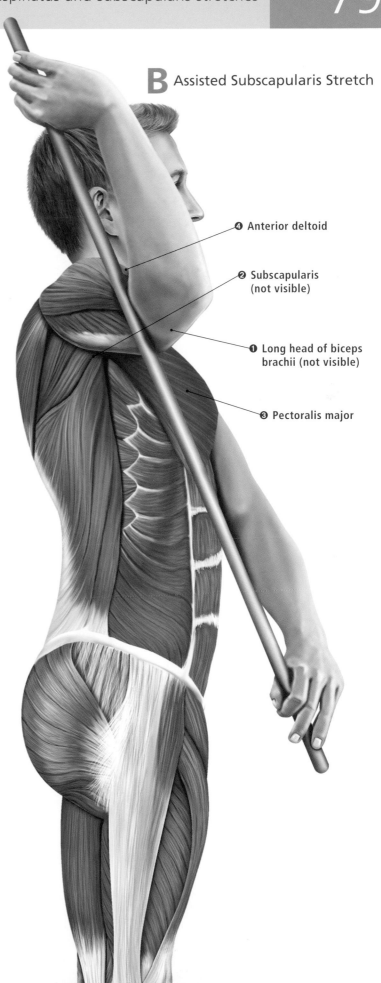

❹ Anterior deltoid

❷ Subscapularis (not visible)

❶ Long head of biceps brachii (not visible)

❸ Pectoralis major

Wrap Around and One-arm Abductor Stretches

Most of our daily activities require us to use our hands and arms in front of the body. This, along with the pull of gravity, puts pressure on the muscles across the top and back of our shoulders. The wrap around stretch targets the muscles of the upper back that stabilize the shoulder blades, including trapezius and rhomboids, as well as the muscles that generate posterior movement at the shoulder, including the rear deltoid. The one-arm abductor stretch targets the two muscles that produce abduction, or lifting the arm up to the side. The deltoid generates all the power for this action, while the much smaller supraspinatus initiates the movement and provides stability. These stretches can be used as part of a shoulder strengthening program and can be performed without any equipment.

warning

As injury to the supraspinatus tendon is fairly common, seek advice from a health professional before performing these stretches if you have had any previous shoulder pain.

how to

You can perform the wrap around stretch in either a standing or seated position. Begin by raising your arms to shoulder height, then wrap both arms around the front of your upper body as though each hand is trying to scratch an itch at the back of the other shoulder. Walk your fingers along your shoulders to gradually increase the stretch.

For the one-arm abductor stretch, place the back of your hand against the small of your back, keeping your elbow close to your body. Use your other hand to grasp your wrist and gently pull your arm down and across the back of your body. Be sure to maintain an upright posture.

variations

ONE A To focus on the trapezius and rhomboids, perform the wrap around stretch sitting on the floor with your knees bent and feet flat on the floor. Lean forward and wrap your arms around your knees as though you are giving your knees a hug. Lean back slightly to feel the stretch between your shoulder blades.

ONE B If you have trouble joining your hands behind your back in the one-arm abductor stretch, you can use a towel, gripping it in both hands to bridge the gap.

muscles being stretched

❶ Deltoid
❷ Latissimus dorsi
❸ Teres minor
❹ Infraspinatus
❺ Posterior deltoid
❻ Supraspinatus (under Trapezius)
❼ Trapezius
❽ Rhomboids (under Trapezius)

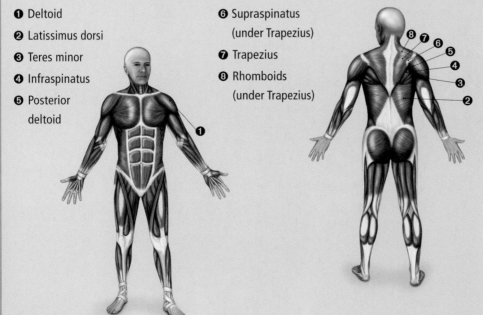

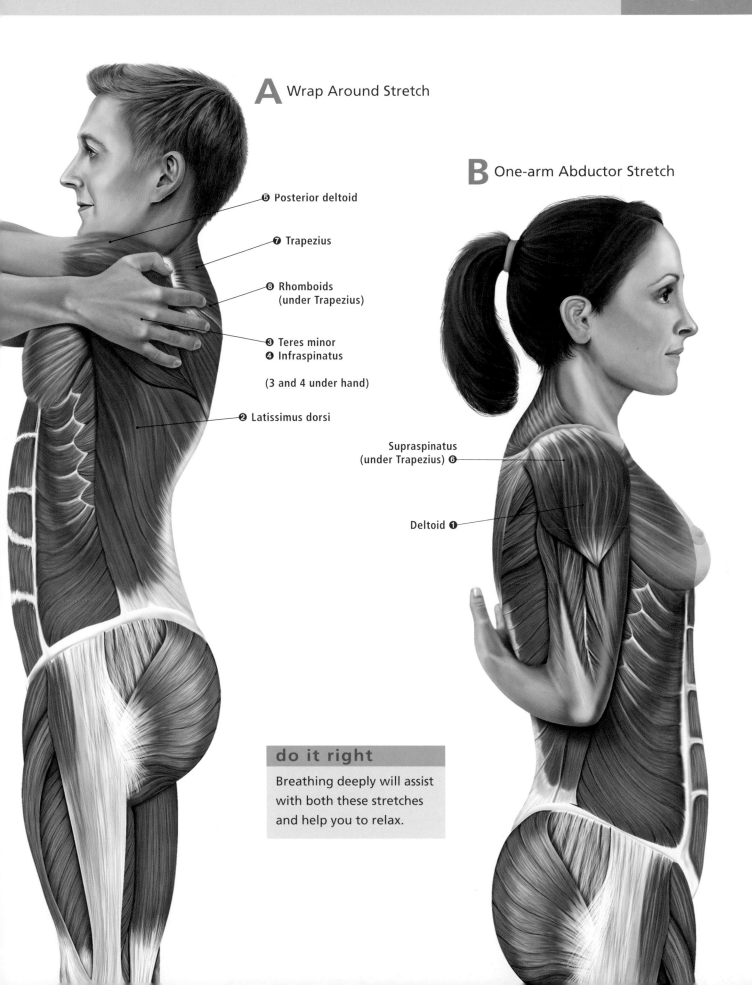

A Wrap Around Stretch

B One-arm Abductor Stretch

❺ Posterior deltoid

❼ Trapezius

❽ Rhomboids
(under Trapezius)

❸ Teres minor
❹ Infraspinatus

(3 and 4 under hand)

❷ Latissimus dorsi

Supraspinatus
(under Trapezius) ❻

Deltoid ❶

do it right

Breathing deeply will assist
with both these stretches
and help you to relax.

Double-arm Abductor Stretch

This stretch targets the workaholic anterior fibers of the deltoid muscle as well as the large and powerful pectoralis major. These muscles represent common areas for tightness in those with poor posture. Excessive tightness in these muscles in conjunction with weakness in posterior muscles may lead to a rounding of the shoulders, back and neck pain, and even difficulty in breathing. This stretch, requiring no equipment, is suitable for most people, and can be done throughout the day to break up prolonged periods sitting at a desk. It is also important for sports, such as rugby or boxing, where you are performing a lot of pushing movements, or for those who do a lot of freestyle swimming.

how to

Place your hands behind your back with your palms facing upward and interlace your fingers. Keeping your elbows straight, your head up, and facing forward, slowly raise your hands upward. Maintain an upright posture and stick your chest out while performing this stretch.

variations

ONE

A If you find it difficult or uncomfortable to join your hands together behind you, you can use a towel to assist with this stretch. Grasp the towel with each hand a short distance apart. Slowly raise your hands upward behind your back. You can adjust the length of towel between each hand until you find what is comfortable for you.

TWO

A You can stretch the left and right anterior deltoid and pectoralis major separately. Find a solid structure that you can grasp behind you such as a piece of gym equipment or even a doorway. Standing upright and facing away from your support, reach behind and hold on to the structure with your thumb facing the floor. Placing one foot forward and one back, gently shift your weight forward until you feel the stretch.

muscles being stretched

❶ Biceps brachii

❷ Coracobrachialis (under Biceps brachii)

❸ Anterior deltoid

❹ Pectoralis major

warning

Keep your shoulders back and avoid the temptation to lean forward when performing this stretch. If you have had a previous shoulder injury, seek advice from your health professional before performing this stretch.

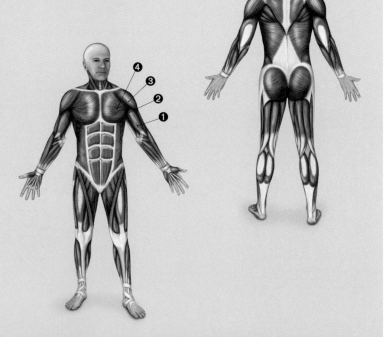

A Double-arm Abductor Stretch

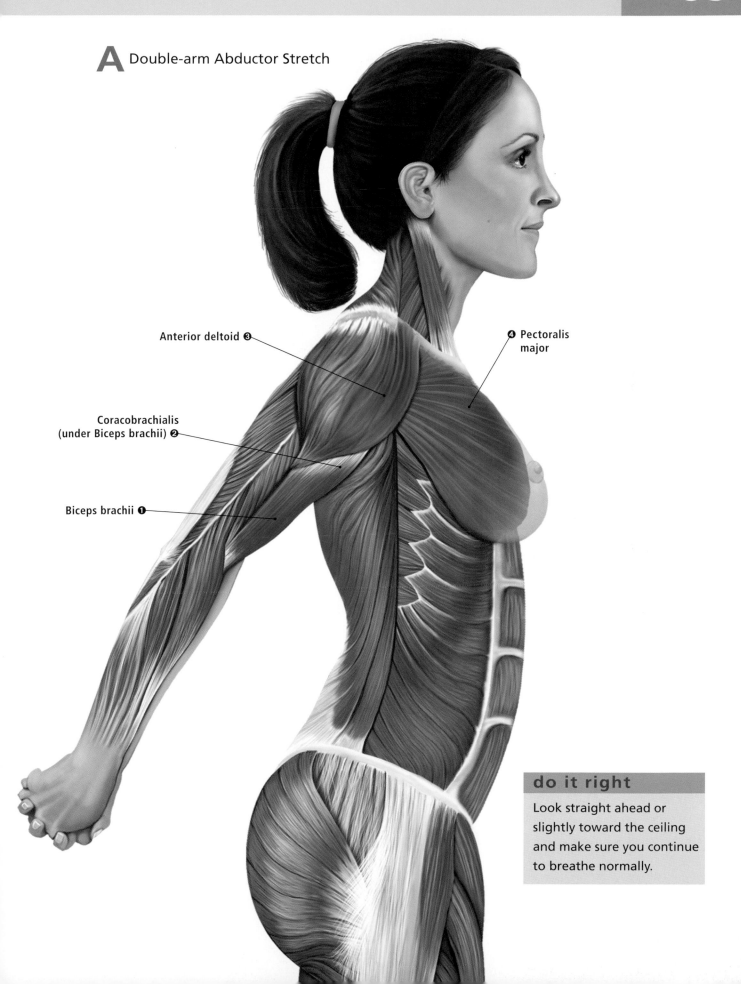

Anterior deltoid ❸

❹ Pectoralis major

Coracobrachialis
(under Biceps brachii) ❷

Biceps brachii ❶

do it right
Look straight ahead or slightly toward the ceiling and make sure you continue to breathe normally.

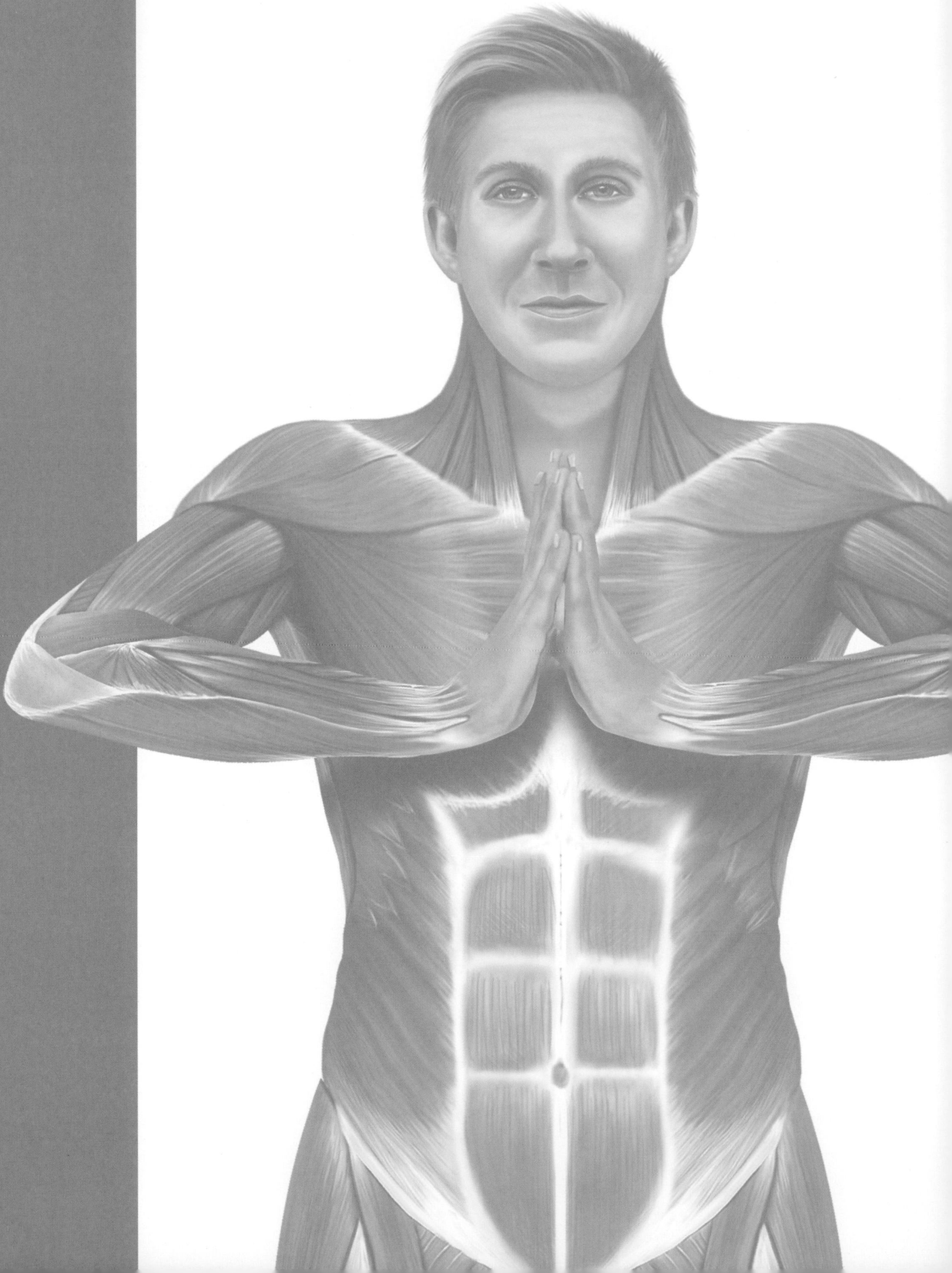

Arm and Forearm Stretches

The muscles of the upper limb produce large and powerful movements at the elbow joint, as well as fine and precisely controlled movements at the fingers. Often, "antagonist" muscles are working in opposite directions to produce more complex movements such as rotation. Prolonged or unaccustomed manual labor or office work can easily result in tightness in the wrists and forearms. Overuse of the wrist flexors or wrist extensors may lead to common injuries such as "golfer's elbow" or "tennis elbow." Taking breaks and including stretching may help to reduce the risk of overuse injuries and other musculoskeletal conditions such as carpel tunnel syndrome.

Joints and Muscles of the Arm and Forearm

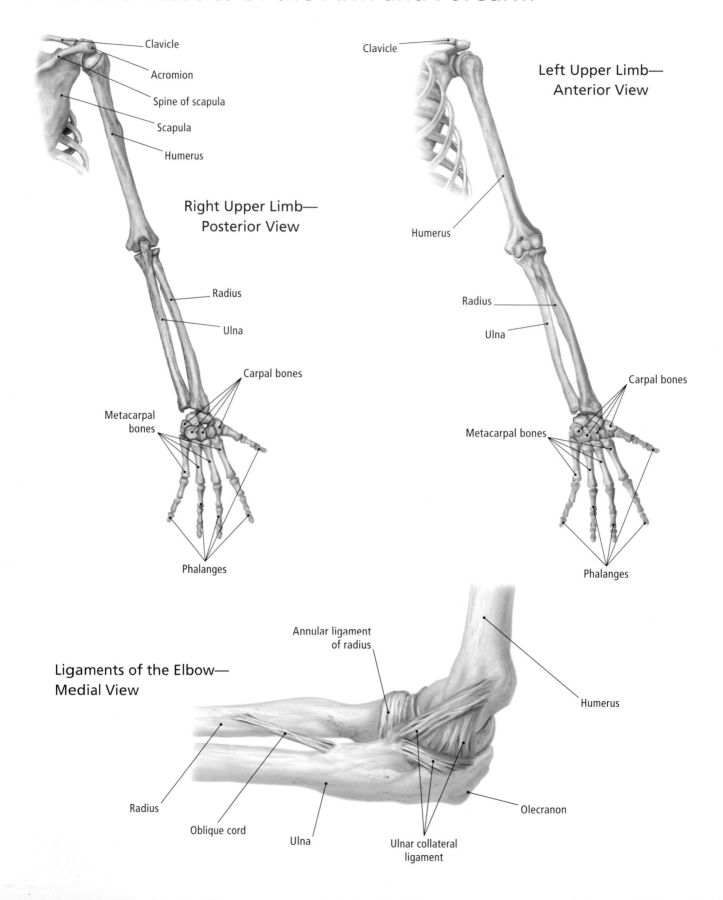

Clavicle

Acromion

Spine of scapula

Scapula

Humerus

Clavicle

**Left Upper Limb—
Anterior View**

**Right Upper Limb—
Posterior View**

Humerus

Radius

Ulna

Radius

Ulna

Carpal bones

Metacarpal bones

Carpal bones

Metacarpal bones

Phalanges

Phalanges

Annular ligament
of radius

**Ligaments of the Elbow—
Medial View**

Humerus

Radius

Oblique cord

Ulna

Ulnar collateral
ligament

Olecranon

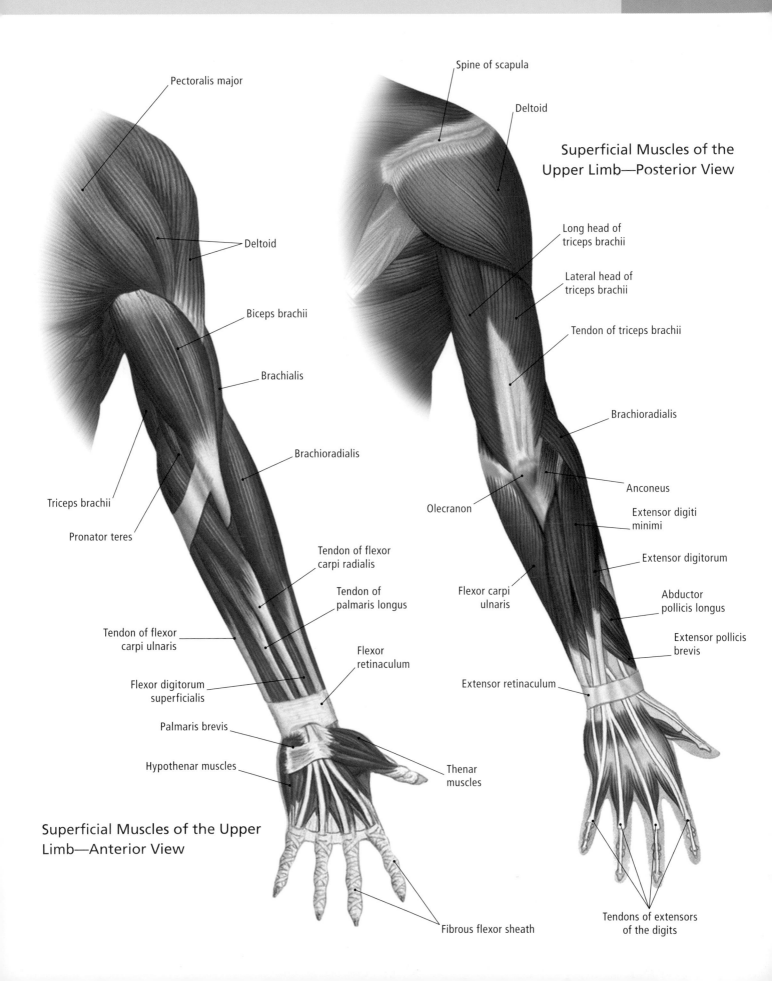

Pectoralis major

Spine of scapula

Deltoid

Superficial Muscles of the Upper Limb—Posterior View

Deltoid

Long head of triceps brachii

Biceps brachii

Lateral head of triceps brachii

Brachialis

Tendon of triceps brachii

Brachioradialis

Brachioradialis

Triceps brachii

Anconeus

Pronator teres

Olecranon

Extensor digiti minimi

Tendon of flexor carpi radialis

Extensor digitorum

Tendon of palmaris longus

Flexor carpi ulnaris

Abductor pollicis longus

Tendon of flexor carpi ulnaris

Extensor pollicis brevis

Flexor digitorum superficialis

Flexor retinaculum

Extensor retinaculum

Palmaris brevis

Hypothenar muscles

Thenar muscles

Superficial Muscles of the Upper Limb—Anterior View

Fibrous flexor sheath

Tendons of extensors of the digits

Triceps Stretch

The triceps are a common site for soreness in those new to a weight-training program. The triceps brachii muscle extends, or straightens, the elbow and is also involved with extending and adducting the shoulder in movements such as a pull-up or a wood-chopping motion. A tight triceps may contribute to ongoing shoulder pain. This stretch effectively brings the muscle into its fully lengthened position, and may be ideal for those involved in activities that call on the triceps brachii such as kayaking, swimming, and weight-training exercises like push-ups or bench-press. The triceps stretch requires no equipment and can be performed while standing or sitting.

warning

Do not pull your arm down on top of your head so that it pushes your neck into flexion.

how to

Raise one arm in the air, then bend your elbow and place your hand on your back between your shoulders. Bring your other hand up and use it to pull your elbow toward the midline and downward until you feel the stretch in the back of your arm.

variations

ONE

A If you are not comfortable reaching over with your other arm, you can use a towel to assist with this stretch. Hold the towel with the arm you wish to stretch. Reach up and place your hand between your shoulder blades so that the towel hangs toward the floor, running down the midline of your back. Grasp the towel with your other hand at the level of your lower back or buttocks, wherever feels comfortable, and gently pull downward.

TWO

A This stretch can be performed so that it involves the latissimus dorsi muscle, which is also used extensively in activities that recruit the triceps brachii such as kayaking and swimming. Raise your arm in the air and place your hand on your back between your shoulders. Bring your other hand up and use it to pull your elbow downward and arch your torso sideways away from the arm that is stretching.

muscles being stretched

❶ Teres major

❷ Triceps brachii

❸ Latissimus dorsi

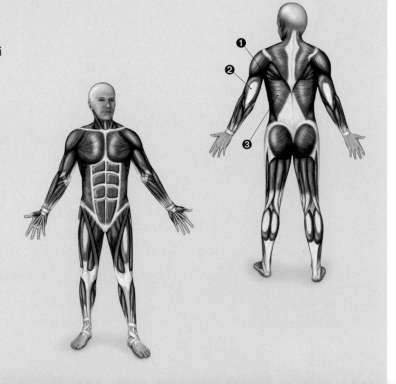

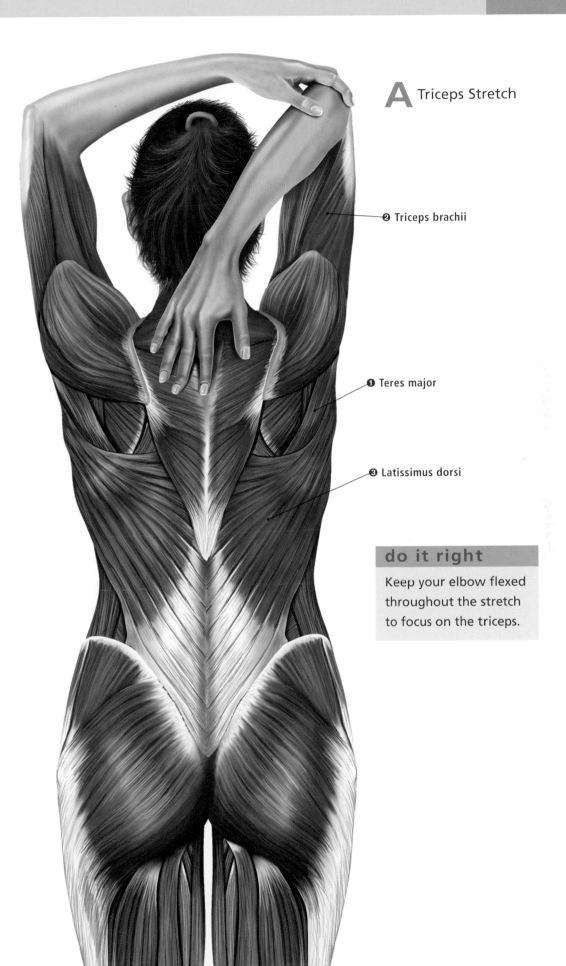

A Triceps Stretch

❷ Triceps brachii

❶ Teres major

❸ Latissimus dorsi

do it right

Keep your elbow flexed throughout the stretch to focus on the triceps.

Kneeling Forearm Stretch

The muscles of the forearm are used to flex the wrist, provide grip strength, and assist with flexing the elbow, as well as to rotate the forearm itself. As such, they are involved in a multitude of different sports and everyday activities. These muscles are implicated in a common overuse injury called "golfer's elbow." This stretch, along with an appropriate strengthening program, can help avoid injury and assist in return to activity after an injury. This stretch can be performed with or without equipment.

warning

Make sure your hands are in a comfortable position before you apply the stretch. Do not try to move them during the stretch.

how to

Start by kneeling on a comfortable surface. Lean forward placing your hands on the floor in front of your knees. Your palms should be flat on the ground with your fingers pointing back toward your knees. In this position, some people will already be able to feel a stretch in the front of their arms. If not, slowly shift your weight by sitting backward, keeping your palms flat on the floor, until you can feel the stretch.

variations

ONE

A For those with limited mobility and flexibility at the wrist, a very similar stretch can be performed that will target most of the wrist flexor muscles. Kneeling on a comfortable surface, place your hands on the floor in front of your knees with your palms flat on the ground and your fingers pointing forward, away from your knees. Slowly shift your weight forward, keeping your palms flat on the floor.

TWO

A If you have difficulty getting up from and down on the floor, you can use a bench or sturdy table to perform this stretch. Choose a table that is a little lower than waist height. Lean forward and place your hands on the table with your palms flat and your fingers pointing back toward you. Apply the stretch by either shifting your weight backward, or by slightly bending your knees.

muscles being stretched

❶ Pronator teres

❷ Brachioradialis

❸ Flexor carpi radialis

❹ Flexor digitorum superficialis

❺ Flexor digitorum profundus (under Flexor digitorum superficialis)

❻ Flexor carpi ulnaris

❼ Palmaris longus

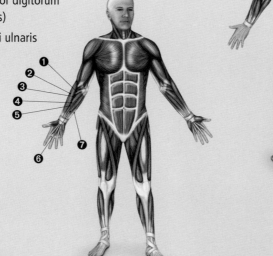
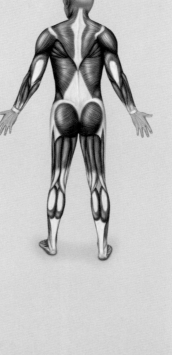

A Kneeling Forearm Stretch

do it right

Keep your elbows straight during this stretch and keep most of your weight on your knees, not your wrists.

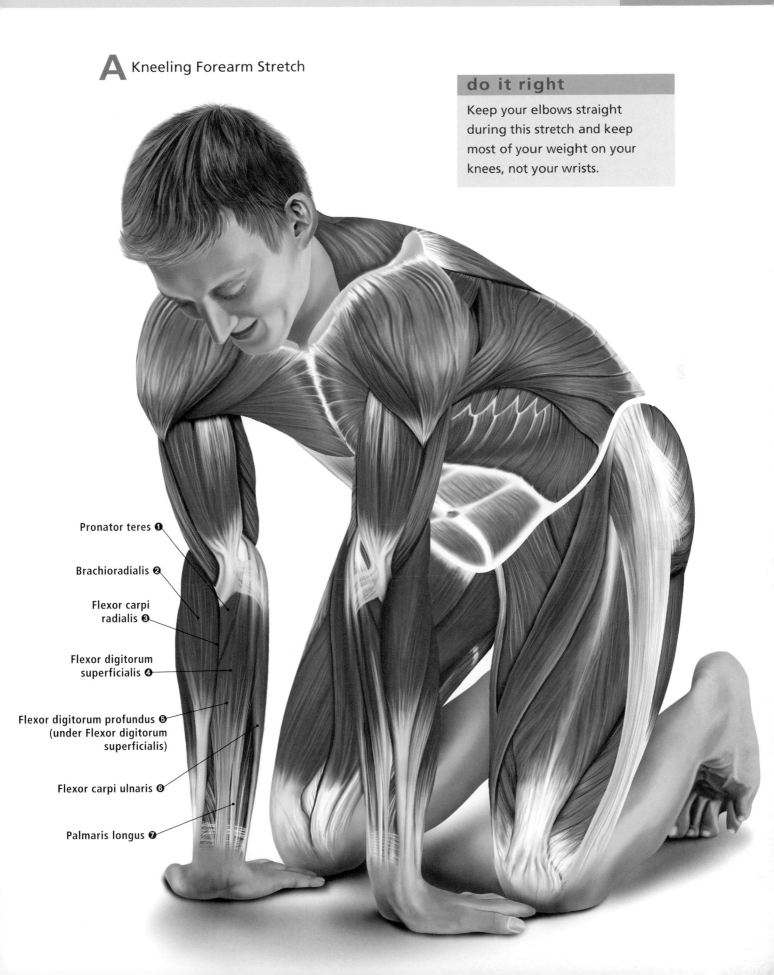

- Pronator teres ❶
- Brachioradialis ❷
- Flexor carpi radialis ❸
- Flexor digitorum superficialis ❹
- Flexor digitorum profundus ❺ (under Flexor digitorum superficialis)
- Flexor carpi ulnaris ❻
- Palmaris longus ❼

Wrist Flexor and Extensor Stretches

These two stretches target the muscles on both the front and the back of the forearm. The muscles on the back of the arm extend the wrist and fingers into a position you would use if you were signaling a person to stop. The muscles on the front of the forearm flex the wrist and close the fingers into a fist. These muscles are particularly important for sports where a high level of grip strength is important such as racquet sports and rock climbing, but they are also essential for a multitude of everyday tasks that require fine control of the fingers, from typing to playing a musical instrument. These stretches can be used after activity, or as a break to relieve tension that has built up throughout the day.

how to

For the wrist flexor stretch, extend your arm in front of you with your elbows straight, palms facing away from you, and your fingers pointing down to the floor. With your other hand, reach forward and gently pull your fingers back. You can angle your arms down toward the ground if you find it uncomfortable to hold them at shoulder height.

For the extensor stretch, hold your arm in front of you with your wrist relaxed so that your hand and fingers fall down toward the ground. Reach forward with your other hand and gently pull your hand and fingers back toward you.

variations

ONE A For a similar stretch, with less emphasis on the brachioradialis, extend your arm in front of you with your elbows straight, palms facing away from you, and your fingers pointing toward the ceiling. Gently pull your fingers back using your other hand.

ONE B The extensor stretch can be performed using a flat surface such as a desk to assist. Hold your arm in front of you with your wrist relaxed so that your hand and fingers fall down toward the desk. Push your arm forward so that your hand and fingers are rolled back toward you by the surface of the desk.

muscles being stretched

❶ Flexor carpi radialis
❷ Extensor carpi radialis longus
❸ Extensor carpi radialis brevis
❹ Flexor digitorum superficialis
❺ Flexor digitorum profundus (under Flexor digitorum superficialis)
❻ Extensor digiti minimi longus
❼ Flexor carpi ulnaris
❽ Palmaris longus
❾ Extensor digitorum
❿ Extensor carpi ulnaris
⓫ Brachioradialis

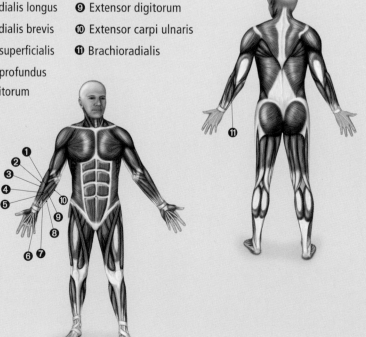

do it right

Apply pressure with the other hand across the whole finger of the stretching hand, not just the fingertips.

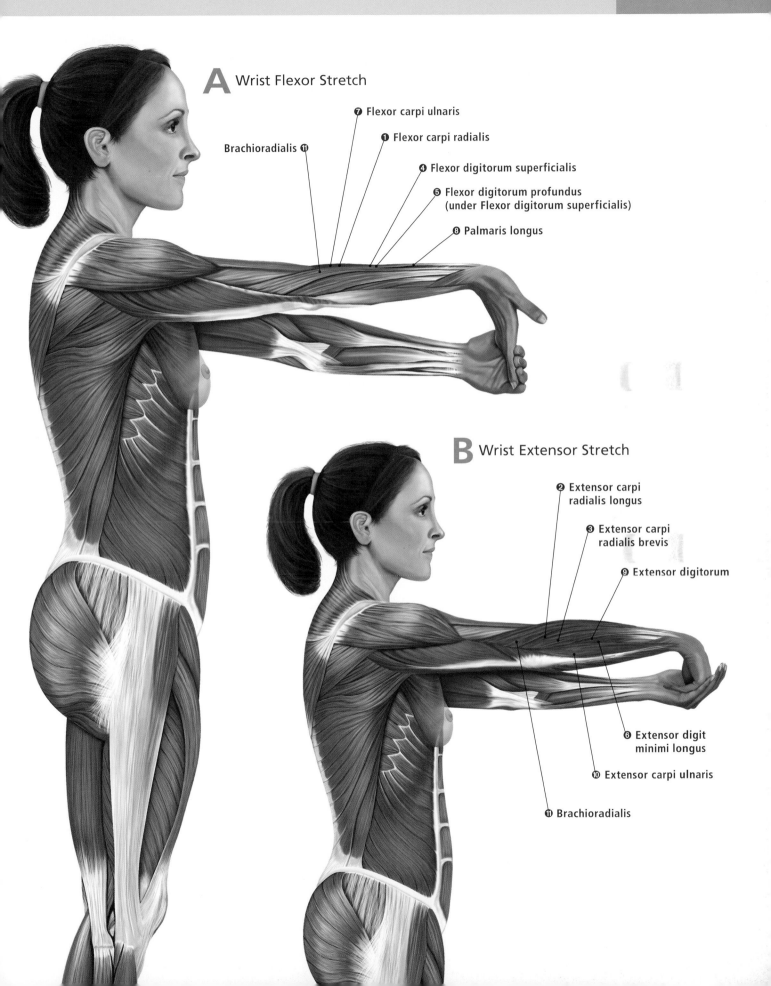

A Wrist Flexor Stretch

❼ Flexor carpi ulnaris

❶ Flexor carpi radialis

Brachioradialis ⑪

❹ Flexor digitorum superficialis

❺ Flexor digitorum profundus
(under Flexor digitorum superficialis)

❽ Palmaris longus

B Wrist Extensor Stretch

❷ Extensor carpi
radialis longus

❸ Extensor carpi
radialis brevis

❾ Extensor digitorum

❻ Extensor digit
minimi longus

⑩ Extensor carpi ulnaris

⑪ Brachioradialis

Rotating Wrist Stretch

This stretch (which can be performed with no equipment) targets the muscles that extend the wrist, as well as the muscles that generate supination at the forearm—a movement that rotates the wrist to face the palm toward the ceiling. These muscles are called upon to perform movements such as driving a screw or tightening a lid. Some sporting activities, occupational tasks, and weight-training activities may result in stiffness and tightness in these muscles, particularly if you are unaccustomed to these activities. The goal of this stretch should be to gradually increase the range of motion through gentle stretching performed regularly.

how to

Extend your right arm straight out in front of you at shoulder height with your wrist flexed so that your fingers point straight down at the floor. Continue to rotate your wrist and arm so that your fingers point out to your right side. Reach over the top with your left arm grabbing the fingers of your right hand and gently pull to continue the rotation of your right wrist and arm. Repeat the procedure on the left side to stretch your other arm.

variations

ONE

A If you can't reach your fingers to continue the stretch, you can grasp your right arm just above the wrist with your left hand and assist the stretch by rotating the forearm.

TWO

A If you have a lightweight pole or rod, you can use it to help with the stretch. Grasp the middle of the pole in your right hand and extend your arm in front of you at shoulder height. Rotate your wrist as though you are pouring out a jug of water. Reach down with your left hand and use it to continue the rotation by pushing the pole down and to the right until you feel the stretch.

muscles being stretched

❶ Flexor digitorum superficialis

❷ Palmaris longus

❸ Flexor carpi ulnaris

❹ Flexor digitorum profundus

❺ Flexor carpi radialis

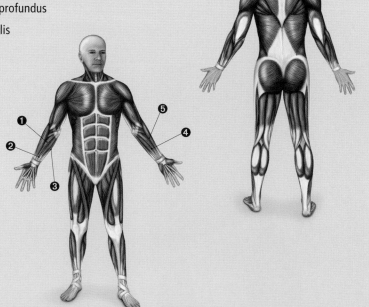

warning

Provide only gentle pressure with the supporting hand: overly vigorous movements may result in injury.

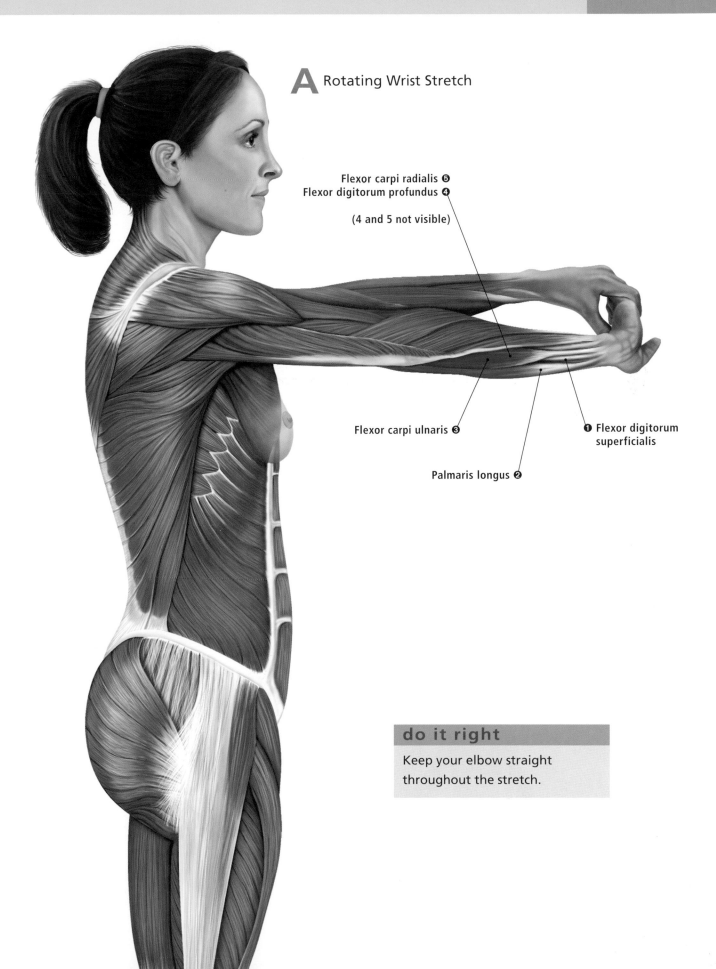

A Rotating Wrist Stretch

Flexor carpi radialis ❺
Flexor digitorum profundus ❹

(4 and 5 not visible)

Flexor carpi ulnaris ❸

Palmaris longus ❷

❶ Flexor digitorum superficialis

do it right
Keep your elbow straight throughout the stretch.

Thumb and Finger Stretches

Often overlooked as part of a stretching program, the muscles that control the fingers and thumb can become tight and fatigued like any other muscles. People who rely on their fingers have always known this: pianists and professional guitarists perform finger stretches to maintain their dexterity and avoid overuse injury. Although they move the fingers and thumb, the muscles themselves are located back in the hand and forearm, so there is some overlap when performing forearm stretches; however, the two stretches here focus on the smaller muscles of the digits.

how to

To stretch your thumb, raise your arm in front of you as though you are about to shake someone's hand, then bend your elbow to 90 degrees so your fingers are pointing straight up in the air. Use your other hand to grasp your thumb and pull it down toward your forearm. Once in this position, you can also gently pull the thumb outward toward the back of the hand.

To stretch your fingers, bring your hands up close to your body at chest height with your elbows pointed out to the side, place your fingertips together, and press your palms toward each other. Lift your elbows higher to increase the stretch.

variations

ONE A To stretch your fingers and wrist flexors at the same time, perform a "prayer stretch" by placing both palms together, keeping your fingers pointed toward the ceiling. Push your hands downward to increase the stretch.

ONE B Interlace your fingers and straighten both arms in front of you with your palms facing away from you. Gently straighten your wrists slightly to increase the stretch in your fingers.

muscles being stretched

❶ Flexor carpi radialis

❷ Flexor pollicis longus

❸ Palmaris longus

❹ Flexor pollicis brevis

❺ Opponens pollicis

❻ Flexor digitorum superficialis

❼ Flexor digitorum profundus

❽ Flexor carpi ulnaris

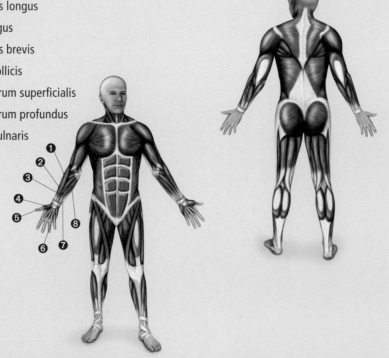

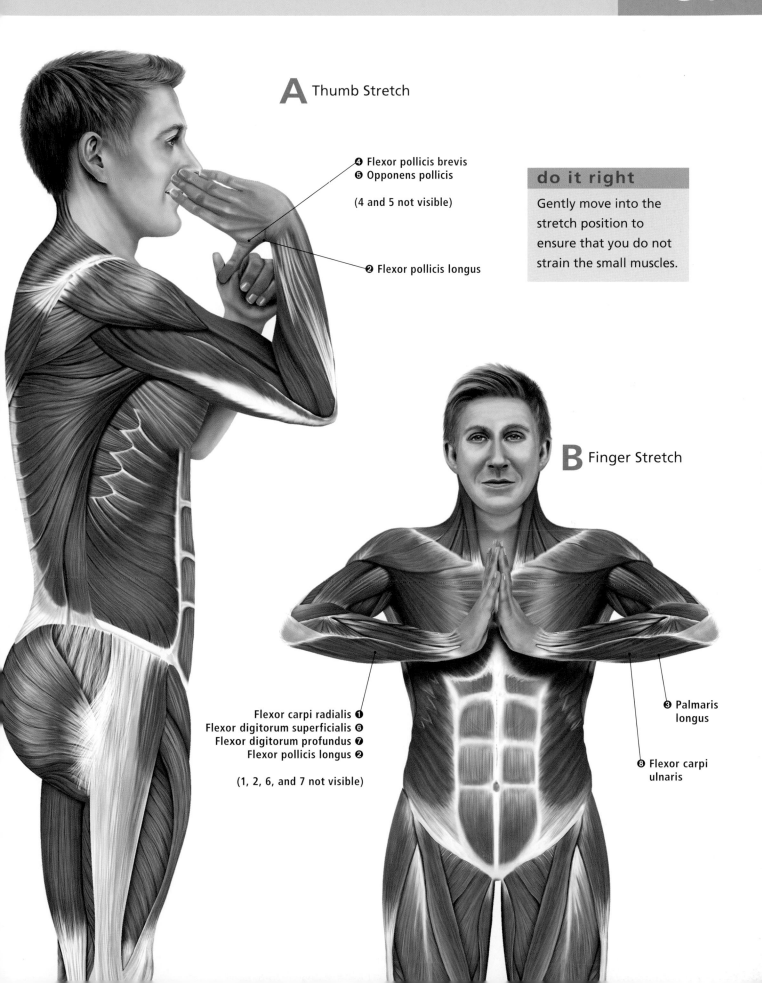

A Thumb Stretch

❹ Flexor pollicis brevis
❺ Opponens pollicis

(4 and 5 not visible)

❷ Flexor pollicis longus

do it right
Gently move into the
stretch position to
ensure that you do not
strain the small muscles.

B Finger Stretch

Flexor carpi radialis ❶
Flexor digitorum superficialis ❻
Flexor digitorum profundus ❼
Flexor pollicis longus ❷

(1, 2, 6, and 7 not visible)

❸ Palmaris
longus

❽ Flexor carpi
ulnaris

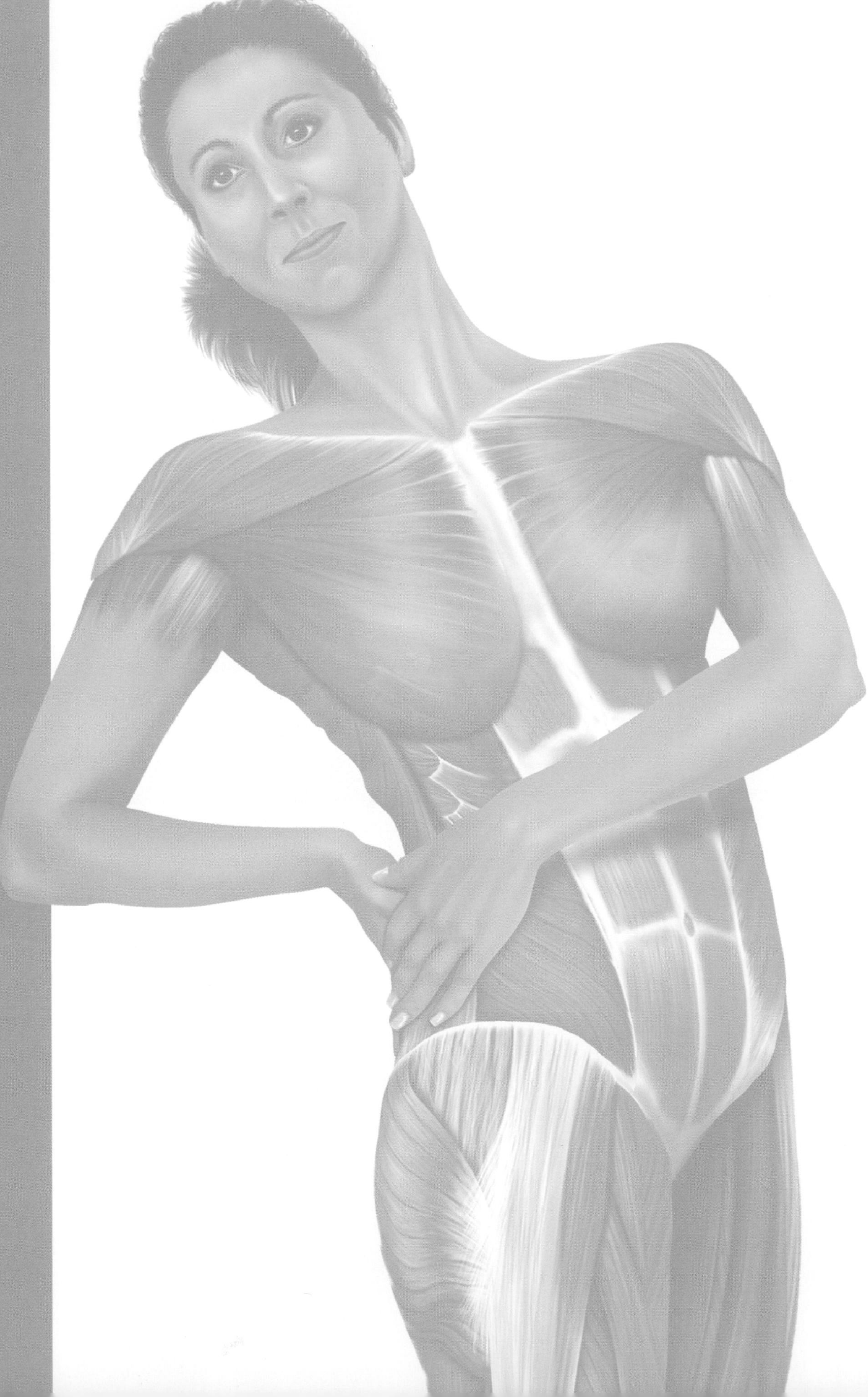

Trunk Stretches

The muscles of the trunk play an important role in providing movement and stability, and they also protect delicate internal organs. Working in union with the muscles of the lower back, these muscles work to flex and extend the torso forward and backward, bend the spine sideways, and rotate the torso. This complex muscle system is also involved in the control of some bodily functions and in assistance with breathing. The trunk muscles can be strained during sporting activities or abdominal exercises; however, maintaining good muscle strength and flexibility may help prevent such strains. You can incorporate these stretches into your exercise routine where the goal is to increase mobility, but you should avoid overstretching this area.

Muscles and Joints of the Trunk

Muscles of the Abdomen—
Anterior View

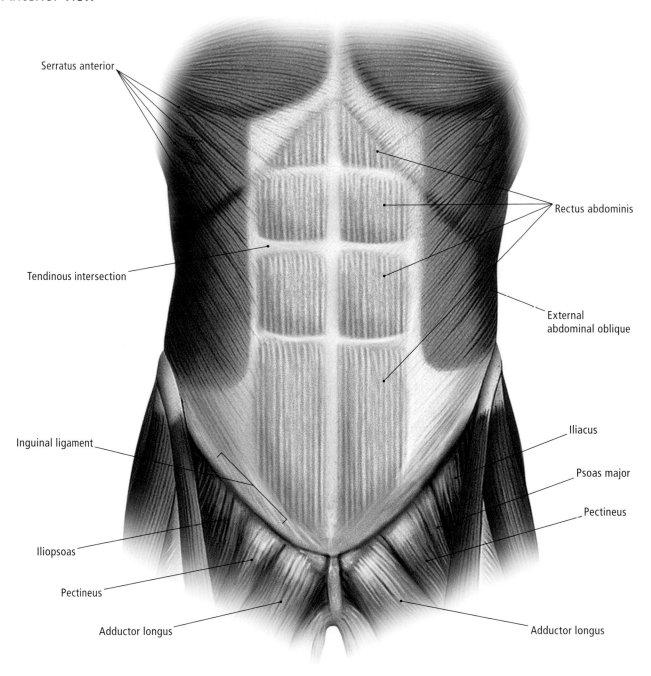

Serratus anterior

Tendinous intersection

Inguinal ligament

Iliopsoas

Pectineus

Adductor longus

Rectus abdominis

External
abdominal oblique

Iliacus

Psoas major

Pectineus

Adductor longus

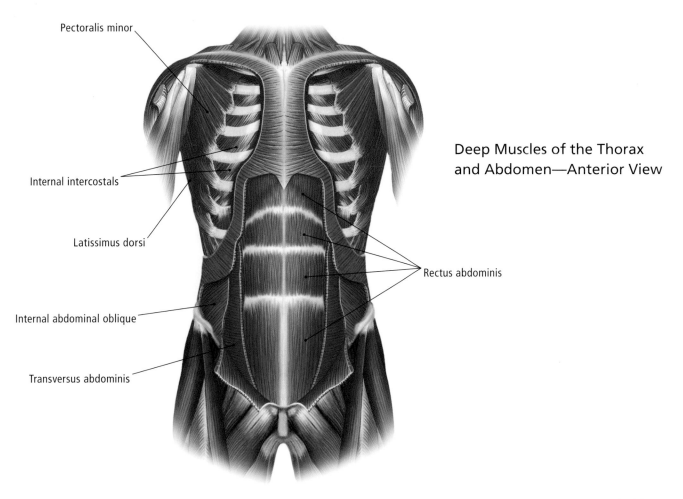

Pectoralis minor

Internal intercostals

Latissimus dorsi

Internal abdominal oblique

Transversus abdominis

Deep Muscles of the Thorax
and Abdomen—Anterior View

Rectus abdominis

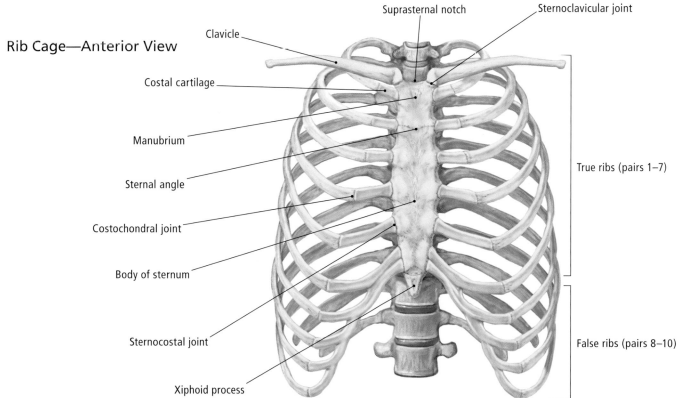

Rib Cage—Anterior View

Suprasternal notch

Sternoclavicular joint

Clavicle

Costal cartilage

Manubrium

Sternal angle

Costochondral joint

Body of sternum

Sternocostal joint

Xiphoid process

True ribs (pairs 1–7)

False ribs (pairs 8–10)

Supported Abdomen and Cobra Stretches

The anterior abdominal muscles, including the well-known and prominent rectus abdominis, are important for the stabilization of the trunk. A tendinous band called the "linea alba" divides the rectus abdominis down the middle, and, along with three more horizontal tendinous intersections, gives the muscle its familiar look. The abdominal muscles are often focused on during an exercise program, with many people performing countless crunches in search of the "six-pack." However, this area is less often stretched, even though these muscles are prone to soreness after exercise, as well as strains, just like any other muscles. These abdominal stretches, in conjunction with a strengthening program that encompasses the whole trunk, will help to improve posture.

how to

For the supported abdomen stretch, start by sitting on an stability ball and slowly walk your feet forward until you are lying supine with the ball in the middle of your back. Ensure you keep your feet wide apart and away from the ball to maintain stability. Allow your upper back and shoulders to curve over the ball so that your head can rest on the ball and your arms hang naturally down to the floor. You should be able to feel the stretch in the abdomen with each breath.

For the cobra stretch, lie face down on the ground with your feet together and toes pointed. Put your palms on the ground near the top of your shoulders. Exhale and press your body up, arching your back while keeping your hips on the ground.

variation

ONE

B If you are new to the cobra stretch, begin by lying face down on the ground with your feet together and toes pointed. Arch your back while keeping your hips on the ground by supporting your upper body with your elbows and forearms on the ground rather than your hands. This will reduce the extension in the spine and the intensity of the stretch.

muscles being stretched

❶ Pectoralis major
❷ Rectus abdominis
❸ External oblique
❹ Rectus femoris
❺ Iliacus
❻ Psoas major

❼ Pectineus
❽ Tensor fasciae latae
❾ Latissimus dorsi
❿ Teres major

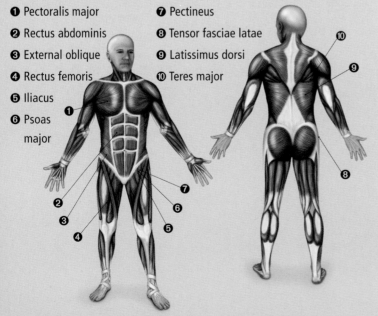

do it right

Until you are confident and balanced, have a partner help you with the supported abdomen stretch in case you lose balance.

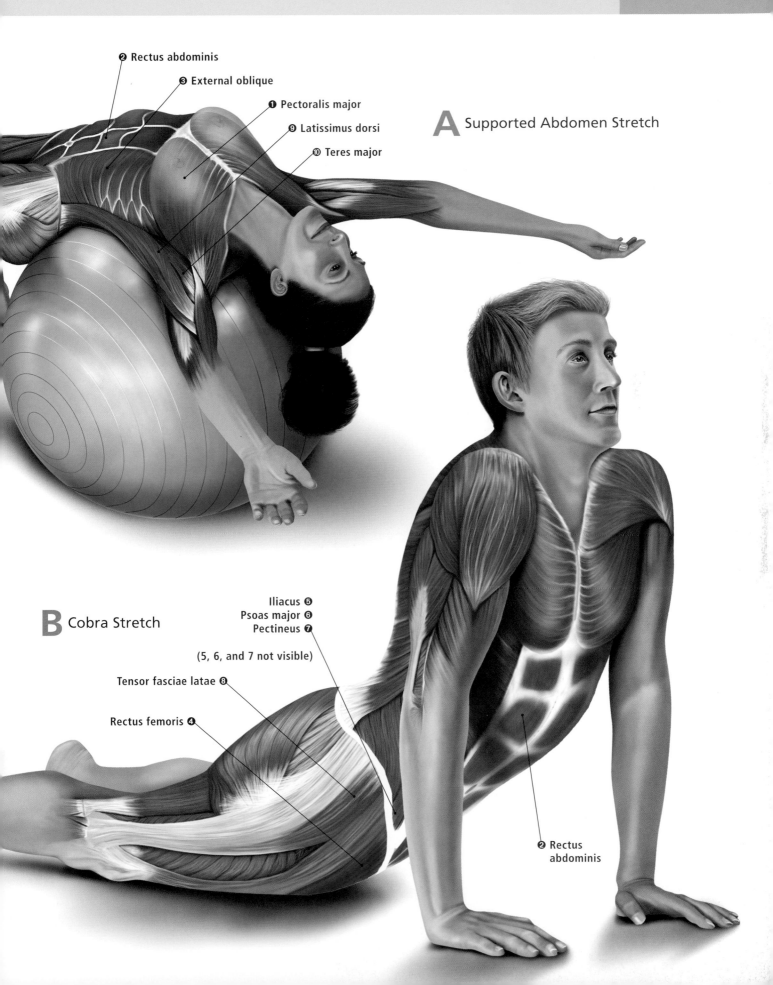

❷ Rectus abdominis
❸ External oblique
❶ Pectoralis major
❾ Latissimus dorsi
❿ Teres major

A Supported Abdomen Stretch

B Cobra Stretch

Iliacus ❺
Psoas major ❻
Pectineus ❼

(5, 6, and 7 not visible)

Tensor fasciae latae ❽

Rectus femoris ❹

❷ Rectus abdominis

Standing Side Stretch

This movement stretches muscles all along the side of the torso. These muscles provide postural support when standing and sitting, as well as generate movements such as side flexion and rotation of the trunk. The obliques assist with trunk flexion and others assist with respiration, particularly during exercise. This stretch, best performed in a standing position, is appropriate for many different sports that involve these movements of the trunk, including golf, swimming, and dance, as well as other activities such as house or yard work. This stretch can also be performed as a part of a warm up where you bend from one side to the other without holding the stretch. Seek advice from a health professional before performing this stretch if you have previously had back problems.

warning

Avoid rotation or bending forward or back at the waist, as this may put additional stress on the spine and cause injury.

how to

Stand with your feet shoulder-width apart or wider. Clasp your hands together in whatever position feels most comfortable for you. Bend your trunk sideways so that your ear begins to point toward the ground. In order to maintain balance, your hips should jut out to the opposite side. Move into position slowly and, while holding the stretch, continue to breathe normally.

variations

ONE

A You can reduce the intensity of the stretch by having only one arm above your head. To stretch your right side, raise your right arm in the air and either place your left hand on your hip or leave it by your side. Arch your trunk sideways, reaching over the top with your right arm.

TWO

A For a gentler stretch, stand with your feet narrower than shoulder-width apart. Keep both hands on your hips and gently bend to one side. Use your hands to support your trunk and limit the stretch to the point you feel comfortable.

muscles being stretched

❶ Rectus abdominis
❷ External and internal obliques

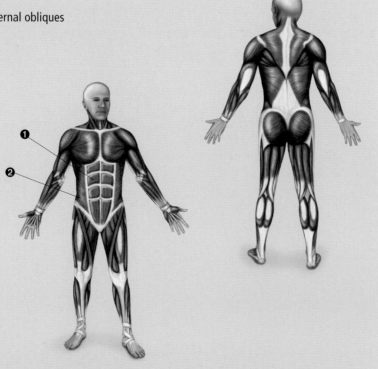

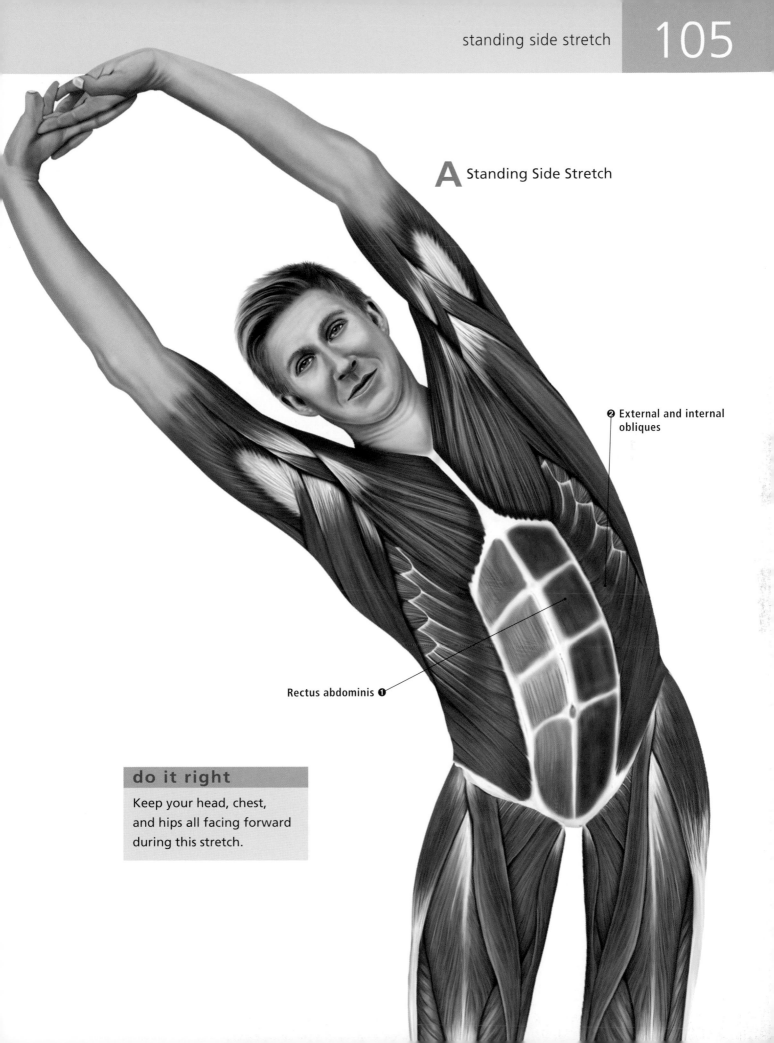

A Standing Side Stretch

❷ **External and internal obliques**

Rectus abdominis ❶

do it right

Keep your head, chest, and hips all facing forward during this stretch.

Kneeling Abdomen and Standing Abdomen Rotation Stretches

The rectus abdominis muscle runs from the bottom of the ribs to the pelvis, and its main action is forward flexion of the spine. In many daily activities, and particularly in sporting movements, the rectus abdominis works in concert with other muscles to generate power. During strong contractions, where a fast movement is required, or against heavy resistance, the oblique muscles join in to assist with trunk flexion. Likewise, the psoas major and rectus abdominis work together synergistically, pulling in opposite directions on the pelvis, to generate forward knee-drive in running and kicking. These stretches target all of these muscles at the same time and will relieve tension after sporting activities or after prolonged sitting.

how to

For the kneeling abdomen stretch, kneel on the floor with your knees hip-width apart and your toes on the ground, not pointed out the back. Place your hands on the upper aspect of your buttocks and arch your spine backward, thrusting your chest out.

For the standing abdomen stretch, stand with your feet hip-width apart and place one hand at the back of your hips. Arch your spine backward and reach around your front with your other arm to meet your other hand, rotating your torso as you do so.

warning

Do not bend sideways at the hips, as this may place extra pressure on the spine.

variations

ONE A For a more advanced kneeling abdomen stretch, find a sturdy overhead bar, like a chin up bar, which you can still reach while standing. Grasp the bar with both hands and place your feet behind you so that your toes appear to be dragging behind you.

ONE B You can perform the standing abdomen rotation stretch in a kneeling position. Kneel on the ground, knees hip-width apart. Place one hand at the back of your hips. Arch your spine backward and rotate your torso, reaching around your body with your other arm.

muscles being stretched

❶ Rectus abdominis
❷ External and internal obliques
❸ Pectineus
❹ Tensor fasciae latae
❺ Sartorius
❻ Vastus medialis
❼ Rectus femoris

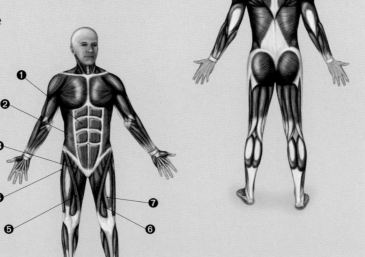

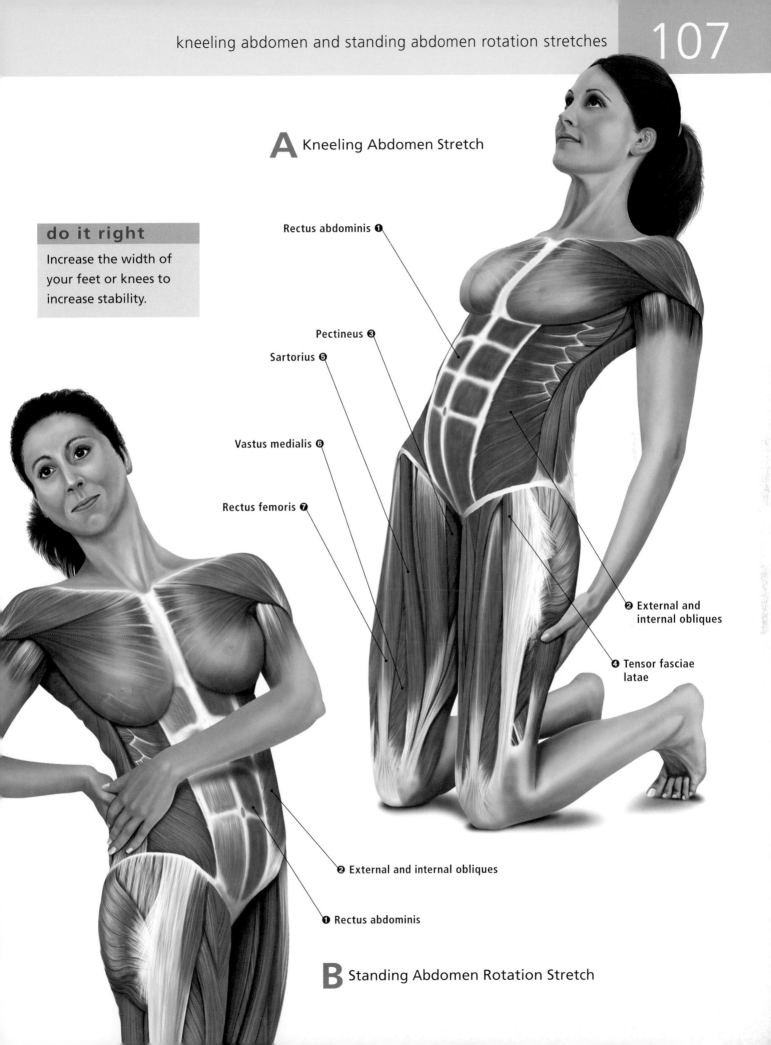

A Kneeling Abdomen Stretch

do it right
Increase the width of your feet or knees to increase stability.

Rectus abdominis ❶

Pectineus ❸

Sartorius ❺

Vastus medialis ❻

Rectus femoris ❼

❷ External and internal obliques

❹ Tensor fasciae latae

❷ External and internal obliques

❶ Rectus abdominis

B Standing Abdomen Rotation Stretch

Back Stretches

The human back incorporates muscles needed to maintain posture and enable a range of movements including pushing, pulling, twisting, throwing, lifting, and climbing. There are large "mover" muscles such as trapezius and latissimus dorsi and small "postural" muscles such as multifidus. The postural muscles work continuously throughout the day while the mover muscles work intermittently depending on the task being performed. Tight muscles in the back can lead to poor posture, difficulty bending and twisting, and poor performance in sporting activities. This in turn can lead to pain or injury. These stretches can be used for general well being, injury rehabilitation, or to achieve specific goals.

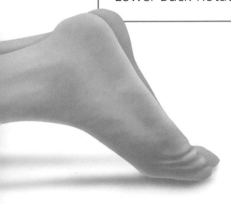

Muscles and Joints of the Back

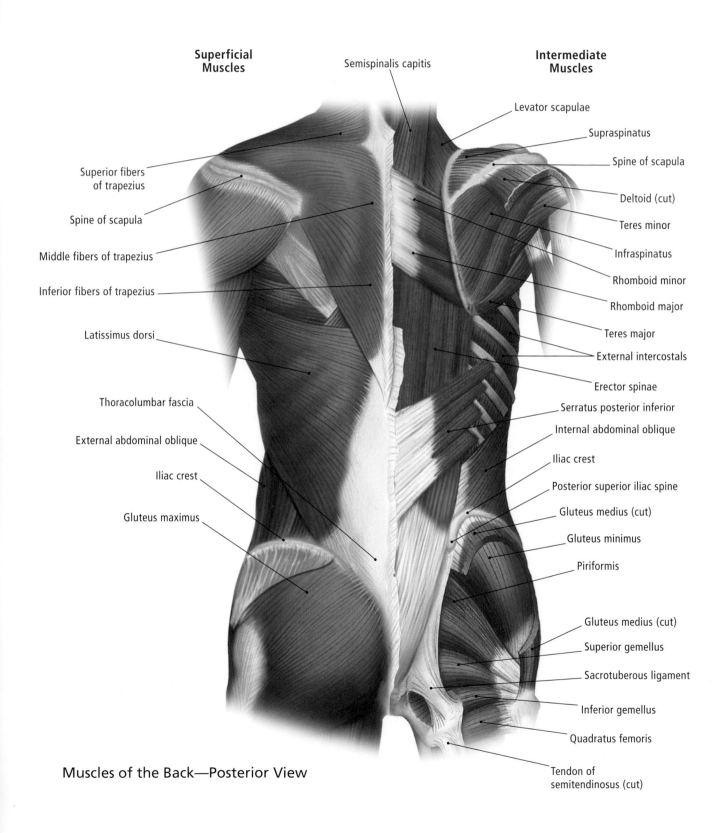

Superficial Muscles

Semispinalis capitis

Intermediate Muscles

Superior fibers of trapezius

Spine of scapula

Middle fibers of trapezius

Inferior fibers of trapezius

Latissimus dorsi

Thoracolumbar fascia

External abdominal oblique

Iliac crest

Gluteus maximus

Levator scapulae

Supraspinatus

Spine of scapula

Deltoid (cut)

Teres minor

Infraspinatus

Rhomboid minor

Rhomboid major

Teres major

External intercostals

Erector spinae

Serratus posterior inferior

Internal abdominal oblique

Iliac crest

Posterior superior iliac spine

Gluteus medius (cut)

Gluteus minimus

Piriformis

Gluteus medius (cut)

Superior gemellus

Sacrotuberous ligament

Inferior gemellus

Quadratus femoris

Tendon of semitendinosus (cut)

Muscles of the Back—Posterior View

Deep Muscles of the Back—Posterior View

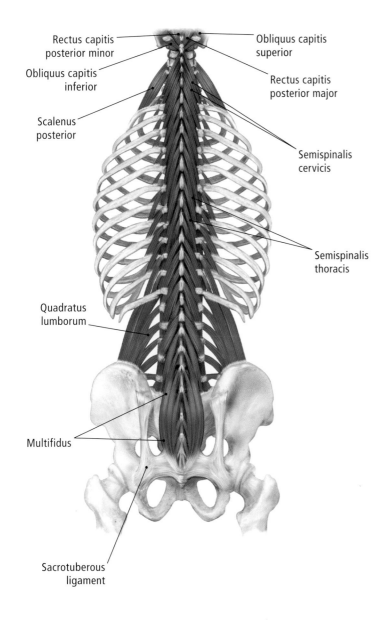

Rectus capitis posterior minor

Obliquus capitis inferior

Scalenus posterior

Obliquus capitis superior

Rectus capitis posterior major

Semispinalis cervicis

Semispinalis thoracis

Quadratus lumborum

Multifidus

Sacrotuberous ligament

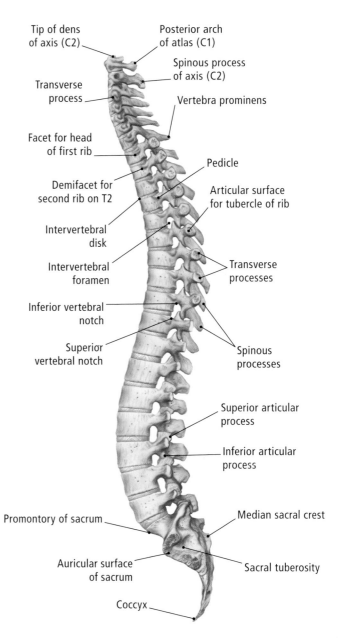

Tip of dens of axis (C2)

Posterior arch of atlas (C1)

Spinous process of axis (C2)

Transverse process

Vertebra prominens

Facet for head of first rib

Pedicle

Demifacet for second rib on T2

Articular surface for tubercle of rib

Intervertebral disk

Intervertebral foramen

Transverse processes

Inferior vertebral notch

Superior vertebral notch

Spinous processes

Superior articular process

Inferior articular process

Promontory of sacrum

Median sacral crest

Auricular surface of sacrum

Sacral tuberosity

Coccyx

Vertebral Column—Lateral View

One-arm Lat Stretch

This standing stretch aims to lengthen the latissimus dorsi (lat). There is a left lat and a right lat. If the stretch is performed by holding with the right hand, the right lat muscle will be stretched, and the left hand stretches the left lat. The latissimus dorsi connects the arm, shoulder blade, and spine. This muscle is used to adduct, extend, and internally rotate the arm; therefore, this could be a useful stretch if any of those movements are limited. For athletes who do a lot of throwing or pulling, the muscle may become tight from fatigue. For this stretch, the only equipment needed is something to hold onto and pull back against.

warning

Make sure the bar that you grasp hold of is secure. Never pull on a door that will open in toward you!

how to

To perform this stretch you need to lean forward, and grasp hold of a bar or pole with the hand in a neutral (thumb up) position with your arm above your head. Lean your body back, keep your chin tucked in toward your chest, and feel the pull from your shoulder and down your side.

variations

ONE

A To increase the intensity of this stretch, turn it into a cross-body stretch. To do this, replicate the stretch as described above, except this time grasp the bar on the opposite side of your body by pronating your hand (thumb down). Take hold of the bar above your left shoulder with your right hand, or right shoulder with left hand.

TWO

A If you are unable to find a bar to hold onto that is below waist height, perform the stretch in a kneeling position rather than standing.

muscles being stretched

❶ External and internal abdominal obliques

❷ Deltoid

❸ Trapezius

❹ Infraspinatus

❺ Teres minor

❻ Teres major

❼ Latissimus dorsi

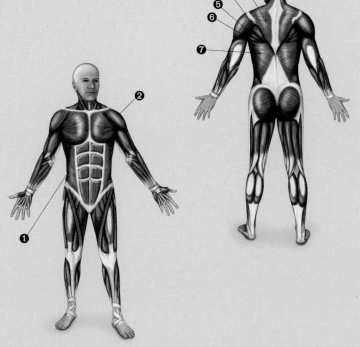

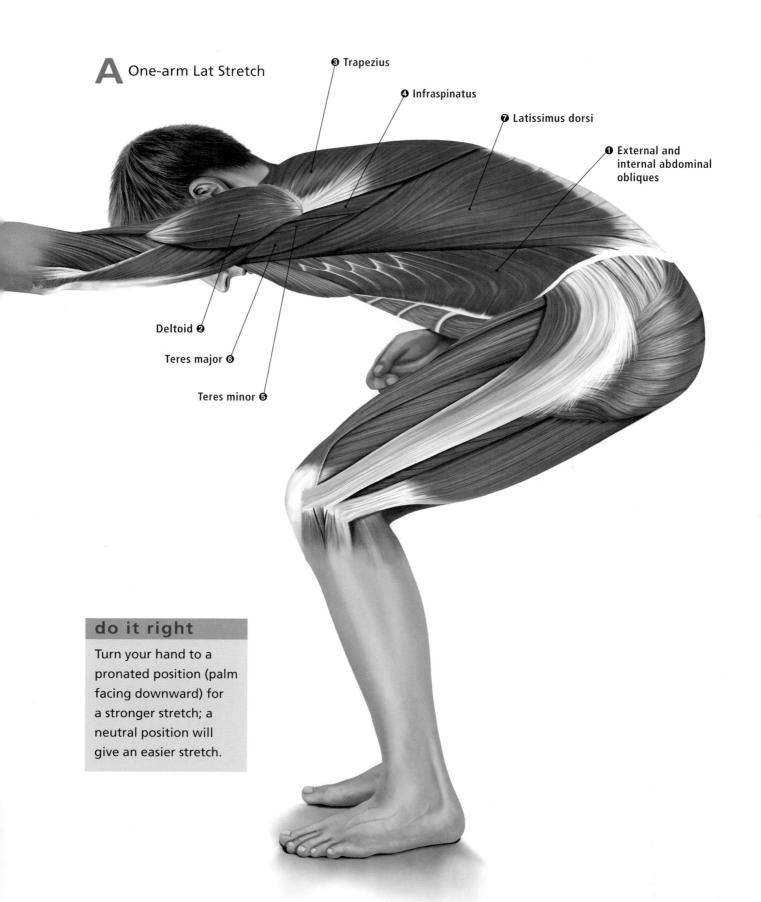

A One-arm Lat Stretch

❸ Trapezius

❹ Infraspinatus

❼ Latissimus dorsi

❶ External and internal abdominal obliques

Deltoid ❷

Teres major ❻

Teres minor ❺

do it right

Turn your hand to a pronated position (palm facing downward) for a stronger stretch; a neutral position will give an easier stretch.

Double Knee-to-chest Stretch and Child Pose

These stretches will lengthen the long muscles of your back—the erector spinae—especially in the lower back, as well as the thoracolumbar fascia. These stretches can also be used to stretch the gluteal muscles of the posterior hip. They are good general back stretches and are often used to help relieve acute back strains. Both of these stretches aim to improve hip and lumbar flexion. Flexion is required when bending, sitting, throwing, or lifting from below waist height. Poor hip and lumbar flexion will result in increased strain being placed on joints of the back and posterior leg muscles such as the hamstrings.

how to

The double knee-to-chest stretch is performed while lying on your back. Bend both hips and knees together, starting with the feet flat on the floor. Slowly lift the feet and grasp your legs just below the knees. Gently hug the knees in toward your chest.

The child pose is a similar stretch that enables greater stretching of the upper and middle back. For this stretch, you need to be able to kneel on the floor, with your feet in a plantar flexed position (top of feet on the floor). Once in this position, lower your hips toward your heels, and stretch forward with your arms above your head.

variations

ONE A Change the double knee-to-chest to a single leg knee-to-chest. This variation places less stress on the lower back and is often better tolerated by people with lower back pain.

ONE B If you are not able to kneel down on the floor, you can perform this stretch in a sitting position while reaching forward onto a table in front of you.

muscles being stretched

❶ Erector spinae (under Thoracolumbar fascia)

❷ Semimembranosus

❸ Semitendinosus

❹ Biceps femoris

❺ Gluteus maximus

❻ Latissimus dorsi

❼ Teres major

❽ Teres minor

❾ Infraspinatus

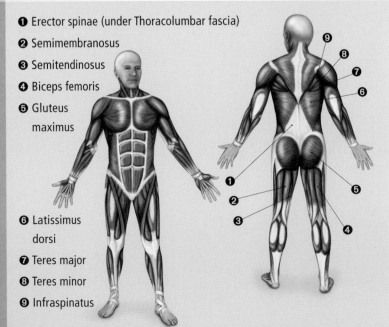

warning

Some back conditions can be exacerbated by flexion stretches.

do it right

Keep your hips down and feel the stretch through your lower back.

A Double Knee-to-chest Stretch

❹ Biceps femoris

❷ Semimembranosus
❸ Semitendinosus

(2 and 3 on inner side of thigh)

❺ Gluteus maximus

❶ Erector spinae (under Thoracolumbar fascia)

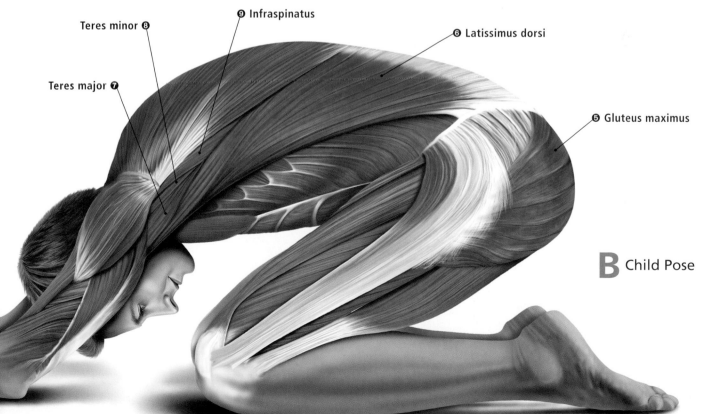

Teres minor ❽

❾ Infraspinatus

❻ Latissimus dorsi

Teres major ❼

❺ Gluteus maximus

B Child Pose

Cow and Cat Stretches

The cow and cat stretches are great for spinal mobility and very good exercises for preventing or relieving back pain in most circumstances. These two stretches are often performed together, moving from one pose to the other. The movement consists of moving from a flexed spinal position to an extended position. It is simple but surprisingly effective at loosening the back, as well as stretching the neck and hips. The movement involves core muscle control and is somewhat beneficial for abdominal strengthening. Many classic yoga routines will include these poses as part of a warm up or for relieving stress and calming the mind.

how to

Start on all fours with your hands under your shoulders and your knees under your hips. The cow stretch is a back extension stretch. Arch your back by moving your belly toward the floor and pushing up through your arms. Lift up your head and look up to the sky. Squeeze your shoulder blades together. Next, move to the cat position. Draw your belly up to your spine, round your shoulders and back, like a cat stretching its back. Drop your head down and tuck your chin in toward your chest. You should feel your hips and the length of your spine mobilizing as well as stretching through the back and posterior shoulder area.

variations

ONE

A + B If you have joint pain, you may be more comfortable performing these poses with your forearms resting on the floor rather than your wrists. You may need to place some padding such as a folded towel or yoga mat under your knees.

TWO

A + B This is a great variation if you want to avoid kneeling on the floor or would like to stretch while you are confined to sitting for long periods (for instance, at work or traveling). Start by sitting upright with your hands flat on your knees. Roll your pelvis forward to arch your back and tilt your head back as far as you are comfortable (cow position), then roll your pelvis back, curl your spine, roll your shoulders forward, and look down to your navel (cat position).

muscles being stretched

❶ External and internal abdominal obliques

❷ Rectus abdominis

❸ Erector spinae (under Thoracolumbar fascia)

❹ Splenius capitis

❺ Splenius cervicis

❻ Semispinalis capitis

❼ Longissimus capitis

❽ Trapezius

❾ Semispinalis thoracis

❿ Longissimus thoracis

⓫ Iliocostalis thoracis

(4, 5, 6, and 7 all under Trapezius)

(9, 10, and 11 all under Latissimus dorsi and Trapezius)

warning

The cow position can aggravate existing back and neck pain, so check with your medical practitioner before you start.

do it right

Inhale with one movement and exhale with the other. Match the movement to your breathing.

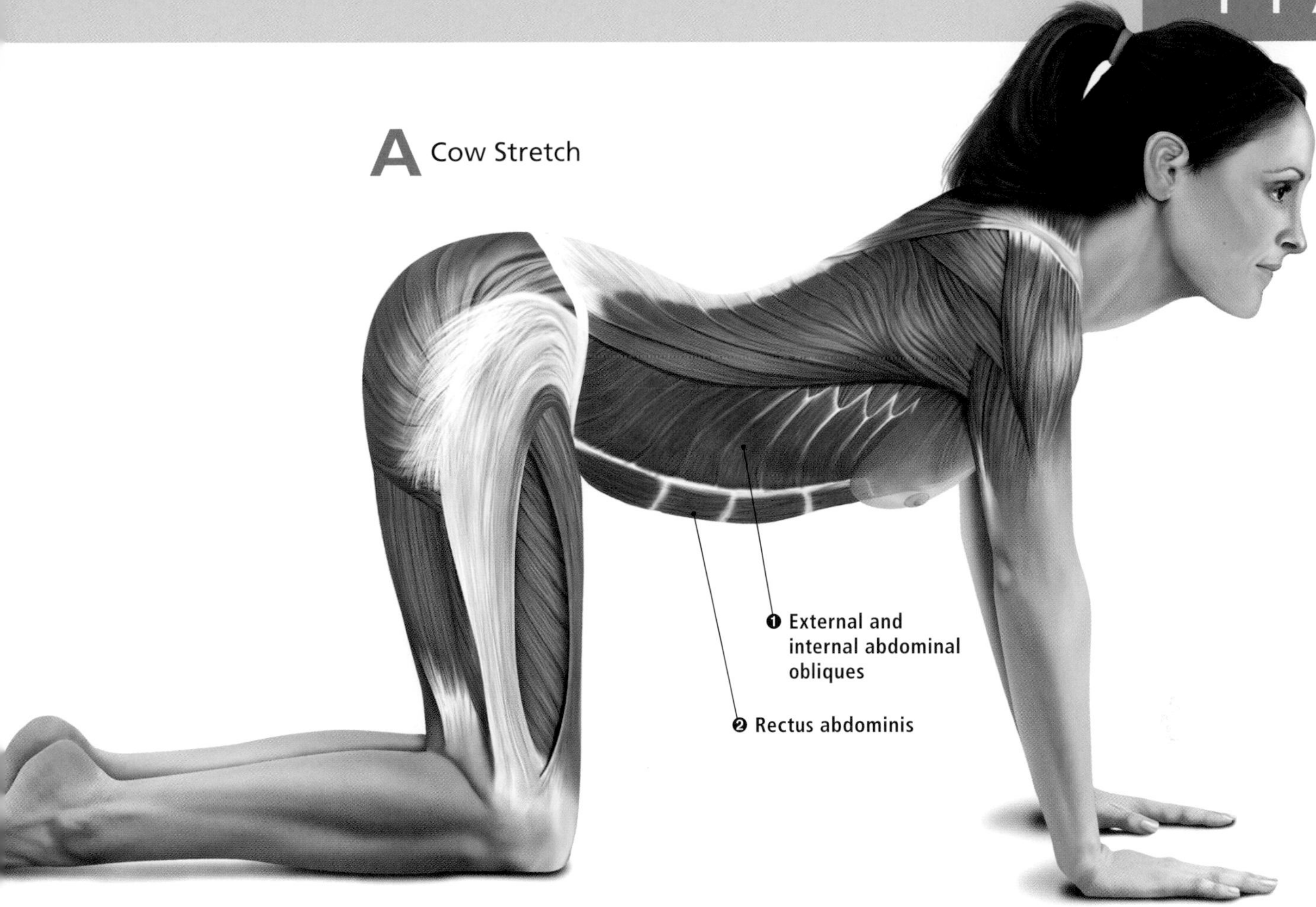

A Cow Stretch

❶ External and internal abdominal obliques

❷ Rectus abdominis

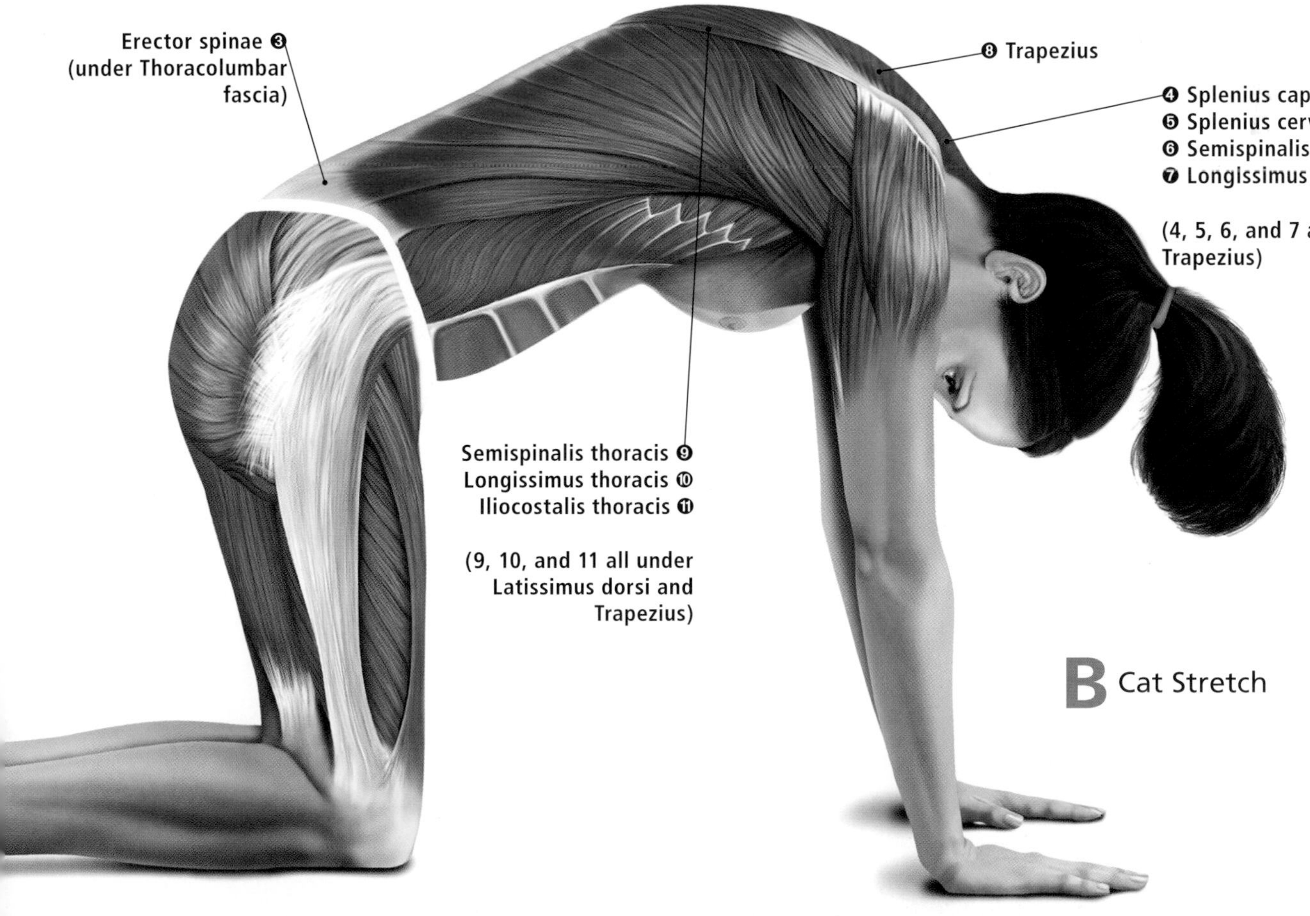

Erector spinae ❸ (under Thoracolumbar fascia)

❽ Trapezius

❹ Splenius capitis
❺ Splenius cervicis
❻ Semispinalis capitis
❼ Longissimus capitis

(4, 5, 6, and 7 all under Trapezius)

Semispinalis thoracis ❾
Longissimus thoracis ❿
Iliocostalis thoracis ⓫

(9, 10, and 11 all under Latissimus dorsi and Trapezius)

B Cat Stretch

Kneeling Back Rotation Stretch

This stretch is particularly useful for opening out the chest and maintaining mobility through the thoracic spine. It is a good stretch to improve posture and relieve the negative effects of sitting all day. Many athletes require trunk rotation to maximize hitting or throwing power. The stretch will help lengthen the pectoralis major, anterior deltoid, rectus abdominis, the abdominal obliques, latissimus dorsi, the gluteals, and the erector spinae group in the back. It can be performed as a sustained stretch or a mobilizing technique. You could add a resistance band to make it a strengthening exercise, or you could balance on one knee while rotating to make it a core stability exercise. As you can see, it is a very versatile position.

how to

Start on the floor on all fours. Keep your chin tucked in toward your chest. Shift your body weight over to one side and slowly lift one hand off the floor. Rotate your spine and stretch out with the arm so that you end up looking up above you and reaching to the sky. Rotate far enough to feel a stretch while still being able to maintain your balance.

variations

ONE

A If you can't comfortably maintain a position on all fours, try the kneeling variation. This stretch is often performed holding a bar across the shoulders. Start this stretch by placing one knee on the ground and the foot of the other leg flat on the ground in front of your knee. Hold the bar across the back of your shoulders and rotate from one side to the other.

TWO

A This common variation is known as the archer stretch. Start this stretch lying on your side with your knees bent together at about 90 degrees. Stretch your bottom arm straight out in front of your shoulder and place the top arm directly on top. Open up through the chest and rotate the spine to take the top arm back across your chest and stretch out behind you, ending up with your shoulders flat and arms stretched out to either side.

muscles being stretched

❶ External and internal abdominal obliques

❷ Upper trapezius

❸ Erector spinae (under Thoracolumbar fascia)

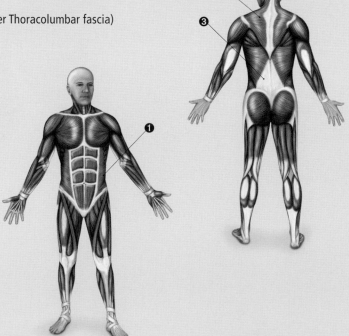

warning

Do not arch your back; always maintain a neutral spine position. If you have any form of back or neck condition, check with your medical practitioner before doing this stretch.

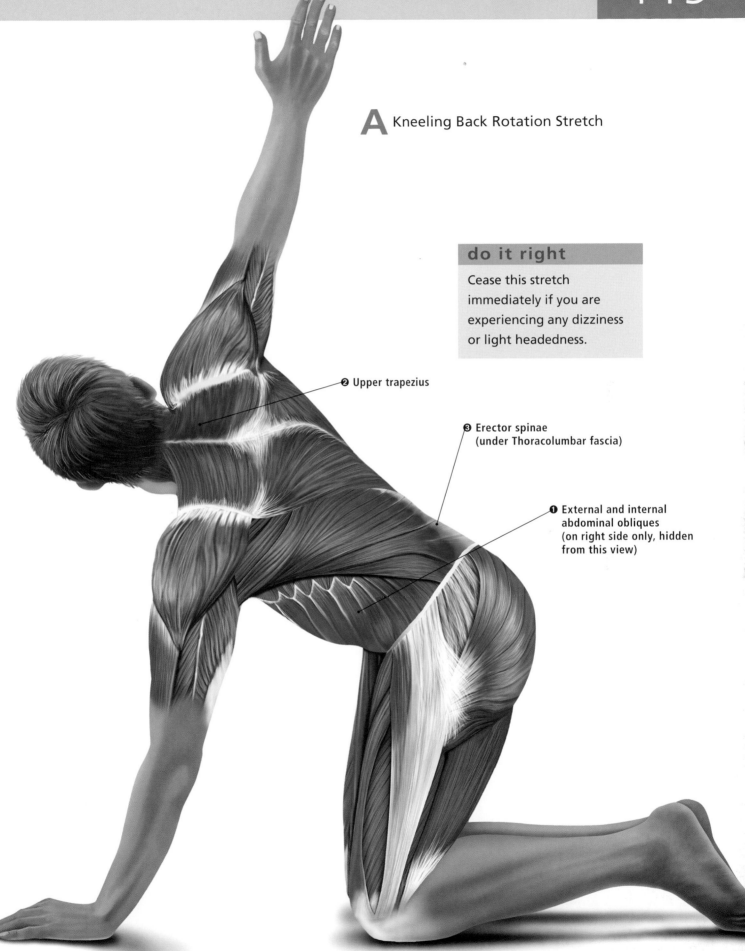

A Kneeling Back Rotation Stretch

do it right

Cease this stretch immediately if you are experiencing any dizziness or light headedness.

❷ Upper trapezius

❸ Erector spinae (under Thoracolumbar fascia)

❶ External and internal abdominal obliques (on right side only, hidden from this view)

Lower Back Rotation Stretch

This is another stretch that is often helpful in relieving acute back strain. It targets muscles from the lower back to the mid back and around the hip. This stretch will improve the ability to rotate the trunk and help reduce lower back muscle tension. People who perform repetitive rotation activities such as driving, assembly line work, packing, loading and unloading, or hitting and throwing sports would be likely to benefit from this stretch. If your work or sport involves repetitive rotation in one direction only, then this stretch will be particularly beneficial when performed in the opposite direction.

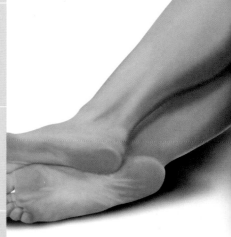

warning

If your head does not rest comfortably on the ground when lying on your back, use a towel or pillow to support underneath. If you have any form of back condition, consult your medical practitioner before doing this stretch.

how to

To perform the stretch, start by lying on your back, with both of your knees bent up to about 90 degrees. Keep your knees and feet together and lay your arms out to the side at around shoulder height with palms facing down. While keeping your knees and feet together, roll your knees to one side as far as your trunk rotation will allow. Return to the start position and then roll your knees to the other side.

variations

ONE

A To increase the difficulty of this stretch, try performing it with your feet in the air. The movement is performed exactly the same way but flex further at the hips to elevate your feet. This turns a simple stretch into a challenging core stability exercise. When performing this variation, ensure that you do not arch your lower back, but instead maintain a "neutral" spine position.

TWO

A To get a stronger and more intense stretch of the back, try performing a single leg rotation. To perform this variation, start with one knee bent and the other leg out straight. Rotate the trunk to lower the bent knee over the top of the straight knee. You can use your hand to pull or hold your knee down (when doing this you must use the left hand for the right leg and vice versa).

muscles being stretched

❶ External abdominal oblique

❷ Rectus abdominis

❸ Erector spinae (under Thoracolumbar fascia)

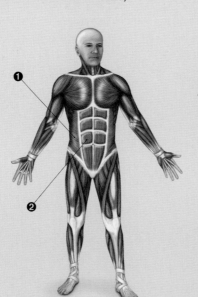

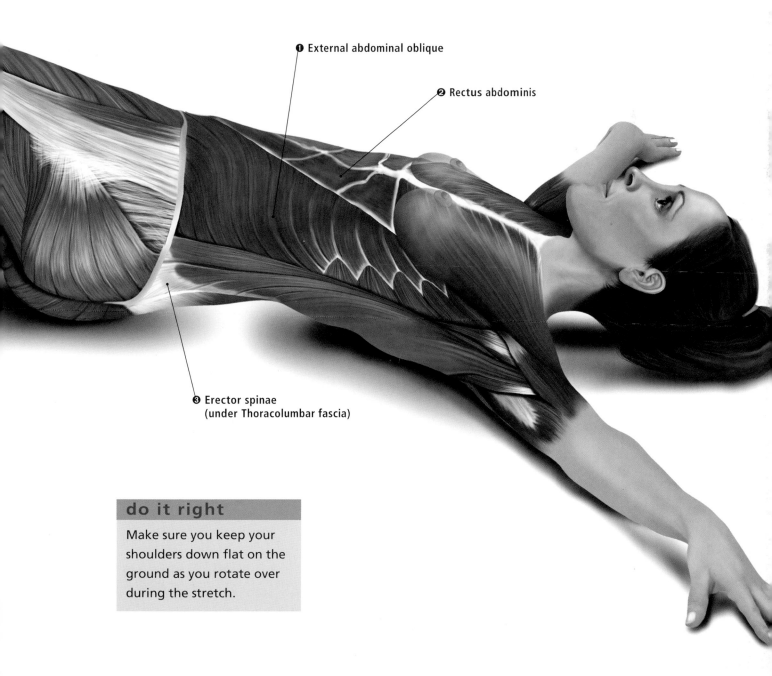

A Lower Back Rotation Stretch

❶ External abdominal oblique

❷ Rectus abdominis

❸ Erector spinae
(under Thoracolumbar fascia)

do it right

Make sure you keep your shoulders down flat on the ground as you rotate over during the stretch.

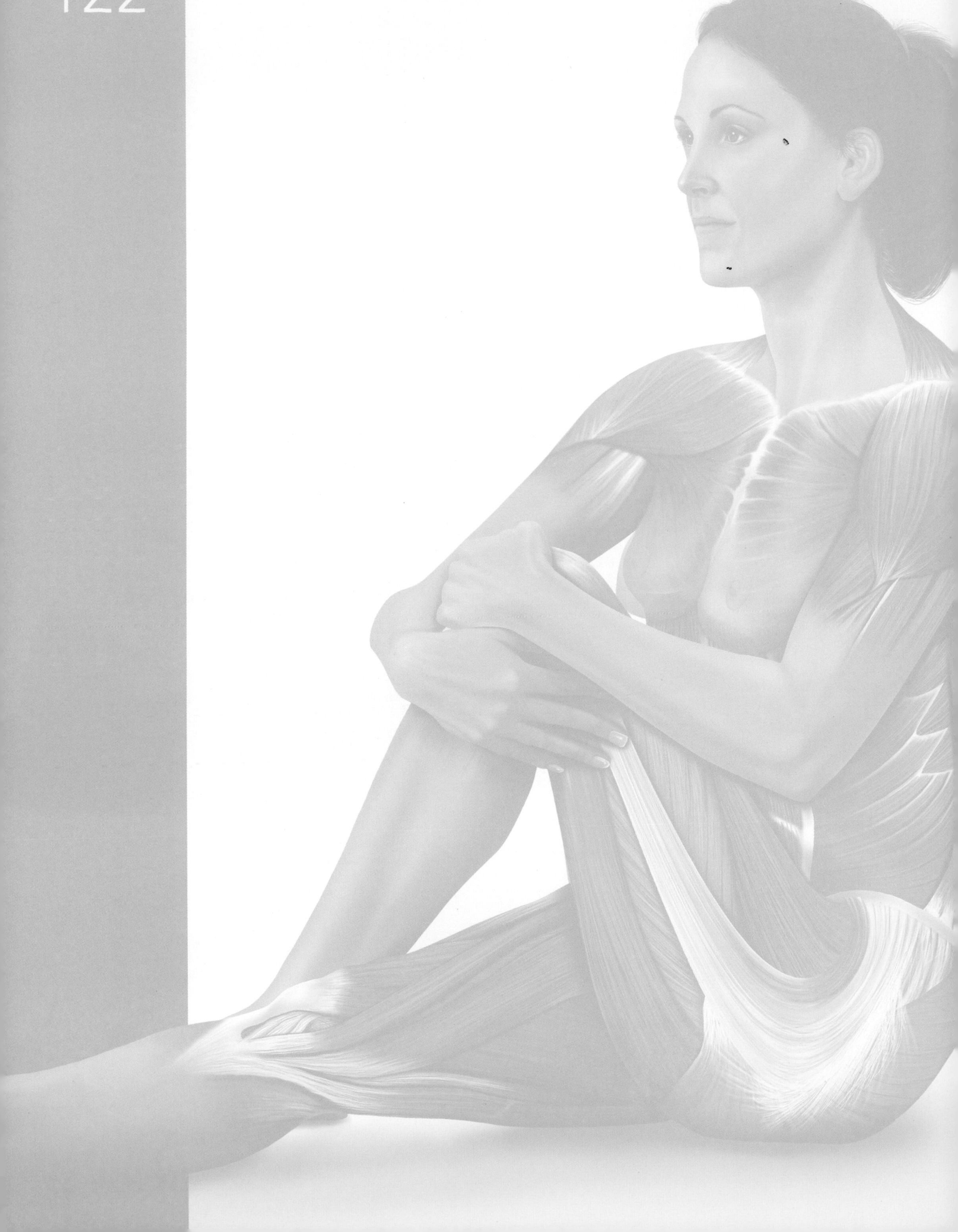

Hip and Buttock Stretches

Muscles around the hip and buttock include the largest muscle in the human body, the gluteus maximus, as well as the gluteus medius and gluteus minimus muscles. There are also deep hip rotators, parts of the hamstring and quadriceps groups, and the long straplike iliotibial band, or tract (ITB). These muscles are particularly active when walking, running, and jumping, and play a significant part in the position of the pelvis, which in turn influences the lordosis and kyphosis (normal curves) of the spine and the centering of the head of the femur into the acetabulum. Common injuries resulting from poor positioning of the pelvis or femur include lower back pain, tendonopathies, hip joint injuries, and knee pain. Tightness in these muscles can occur from extended running or cycling and from prolonged sitting or driving postures.

Muscles and Joints of the Hip and Buttock

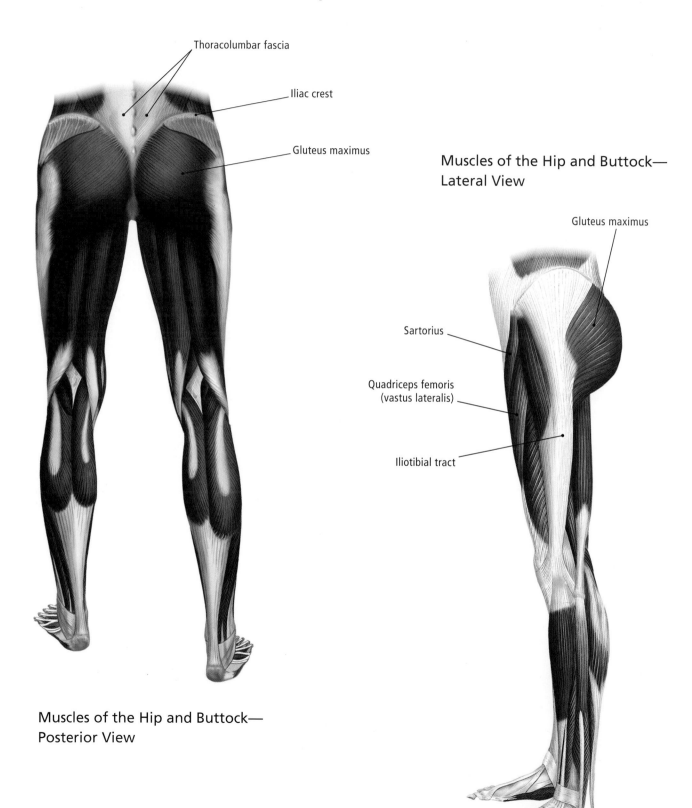

Thoracolumbar fascia

Iliac crest

Gluteus maximus

Muscles of the Hip and Buttock—
Lateral View

Gluteus maximus

Sartorius

Quadriceps femoris
(vastus lateralis)

Iliotibial tract

Muscles of the Hip and Buttock—
Posterior View

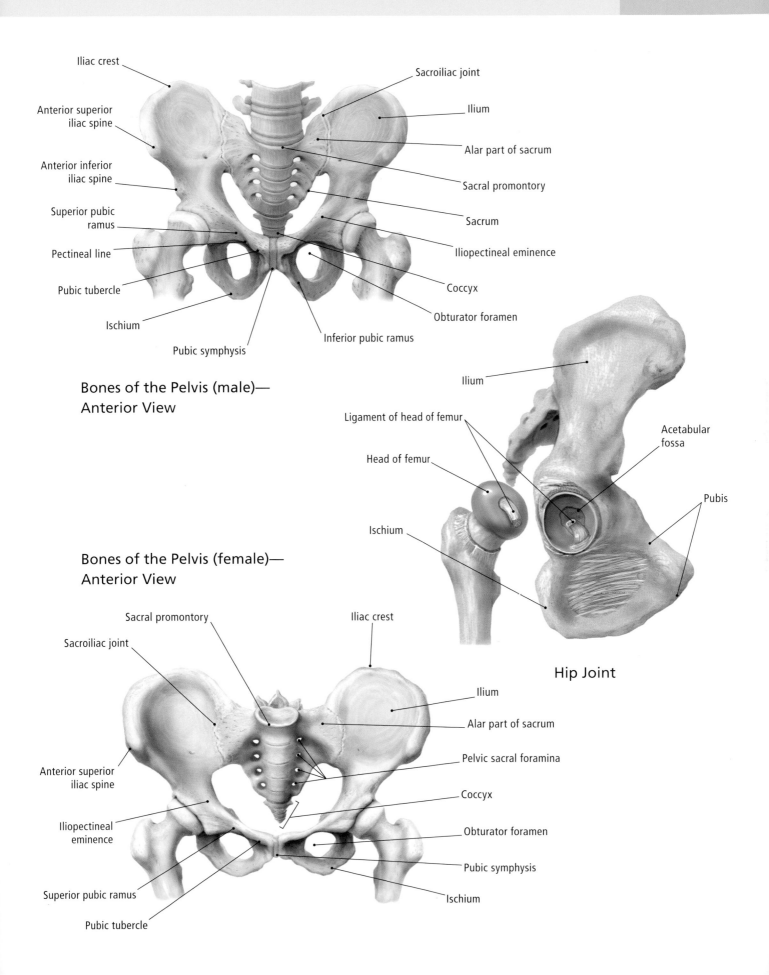

Iliac crest

Anterior superior
iliac spine

Anterior inferior
iliac spine

Superior pubic
ramus

Pectineal line

Pubic tubercle

Ischium

Pubic symphysis

Sacroiliac joint

Ilium

Alar part of sacrum

Sacral promontory

Sacrum

Iliopectineal eminence

Coccyx

Obturator foramen

Inferior pubic ramus

**Bones of the Pelvis (male)—
Anterior View**

Ilium

Ligament of head of femur

Head of femur

Ischium

Acetabular
fossa

Pubis

Hip Joint

**Bones of the Pelvis (female)—
Anterior View**

Sacral promontory

Sacroiliac joint

Iliac crest

Ilium

Alar part of sacrum

Pelvic sacral foramina

Anterior superior
iliac spine

Iliopectineal
eminence

Coccyx

Obturator foramen

Superior pubic ramus

Pubic symphysis

Ischium

Pubic tubercle

Lying Crossover Stretch

This popular stretch is an advanced stretch for the lower back and gluteal region. It will improve trunk rotation, hip external rotation, and hip flexion. This is another stretch that can help relieve acute back strain. For those that find the lower back rotation stretch uncomfortable but want to progress, or those that find the lower back rotation is not providing a great enough stretch, this is likely to be a favorite. You often see this stretch being performed by athletes as part of a dynamic warm up by moving quickly from one side to the other rather than holding a static stretch.

warning

Remember to breathe! This stretch is quite a strain, and some people find they hold their breath and fail to relax.

how to

To perform the stretch, start by lying down flat on your back, ensuring that your head, neck, and shoulders are flat on the floor. Reach your arms out to the side at about shoulder height, with your hand in a supinated (palm up) position. Bend one knee up toward your chest. Then, with the opposite hand placed on the outside of the bent knee, gently pull the knee across your body and toward the floor. Keep your other arm down flat on the floor. Pull your knee far enough over to feel the stretch through your hips and back.

variations

ONE

A Try a "scorpion" stretch by rolling over onto your front side and performing a similar movement. Bend your knee and twist your body so that one leg rotates over the top of the other one. This is another popular dynamic warm up often used by football or soccer teams.

TWO

A For a quick stretch in the office or while out on a walk, try performing this stretch while standing. Lift one knee and pull it across your body in the same way as described above. Just make sure you have good balance and aren't going to fall over!

muscles being stretched

❶ Piriformis

❷ Obturator internus

❸ Obturator externus

❹ Superior gemellus

❺ Inferior gemellus

❻ Quadratus femoris

❼ Gluteus maximus

(1 to 6 all under Gluteus maximus)

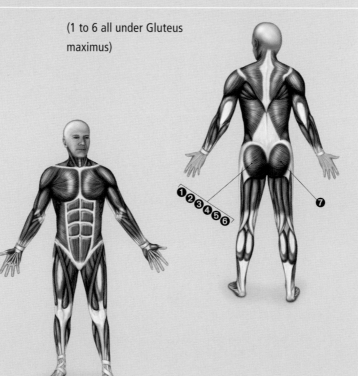
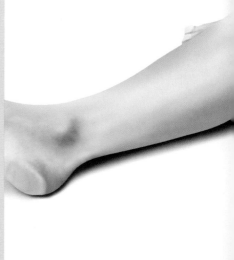

do it right
Make sure you keep your shoulders down flat on the ground as you rotate over during the stretch.

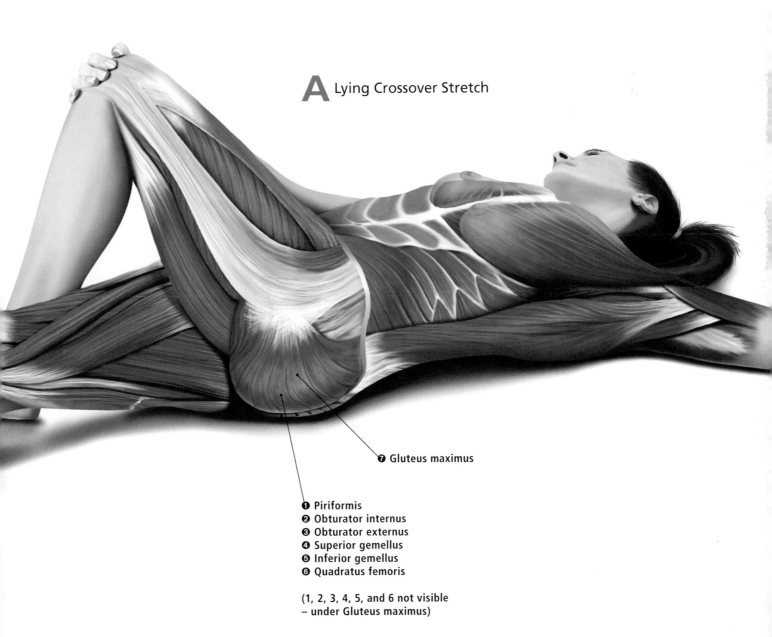

A Lying Crossover Stretch

❼ Gluteus maximus

❶ Piriformis
❷ Obturator internus
❸ Obturator externus
❹ Superior gemellus
❺ Inferior gemellus
❻ Quadratus femoris

(1, 2, 3, 4, 5, and 6 not visible
– under Gluteus maximus)

Sitting Buttock Stretch

This stretch is a favorite among athletes who perform a lot of running in their sports. The main target muscles of this stretch are the piriformis and gluteals muscles, especially the gluteus maximus. Many people like to perform this exercise as part of a lower limb stretching routine following strength workouts that include exercises such as squats and lunges. Despite a widely held belief, stretching does not reduce the incidence of delayed onset muscle soreness (DOMS), but it does reduce tension following a heavy or intense workout. A rare but painful condition known as piriformis syndrome often responds well to stretches such as this.

warning

Do not "bounce" the stretch—hold a static position.

how to

Sit with your legs straight out in front of you (known as long sitting). Bend one leg at the knee and hip to bring the foot alongside the knee of the extended leg. Cross your foot over your straight knee, then use your hands to pull your knee in toward your chest. Keep your back straight and feel the stretch in the back and outer side of your hip.

variations

ONE   **A** Try a lying figure 4 stretch. This time, start by lying on your back with both legs bent up to about 90 degrees. As above, cross one foot over the knee of the opposite leg. Now take hold of the back of the thigh of the leg still on the ground. To do this, one hand has to reach though the hole (or 4) created by the crossed leg. This time, instead of pulling the crossed leg in toward your chest, pull the other leg.

TWO   **A** Add to the stretch by rotating your back as you pull the knee in toward your chest. To do this, start the same way, but instead of grasping the knee and pulling, twist so that you push the knee toward your chest with the back of your opposite elbow.

muscles being stretched

❶ Piriformis
❷ Obturator internus
❸ Superior gemellus
❹ Inferior gemellus
❺ Quadratus femoris
❻ Gluteus maximus

(1 to 5 all under Gluteus maximus)

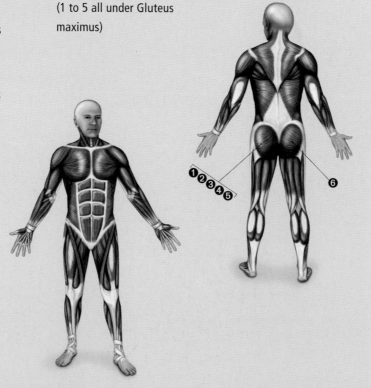

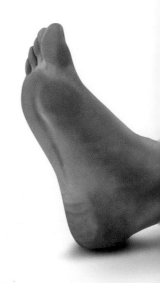

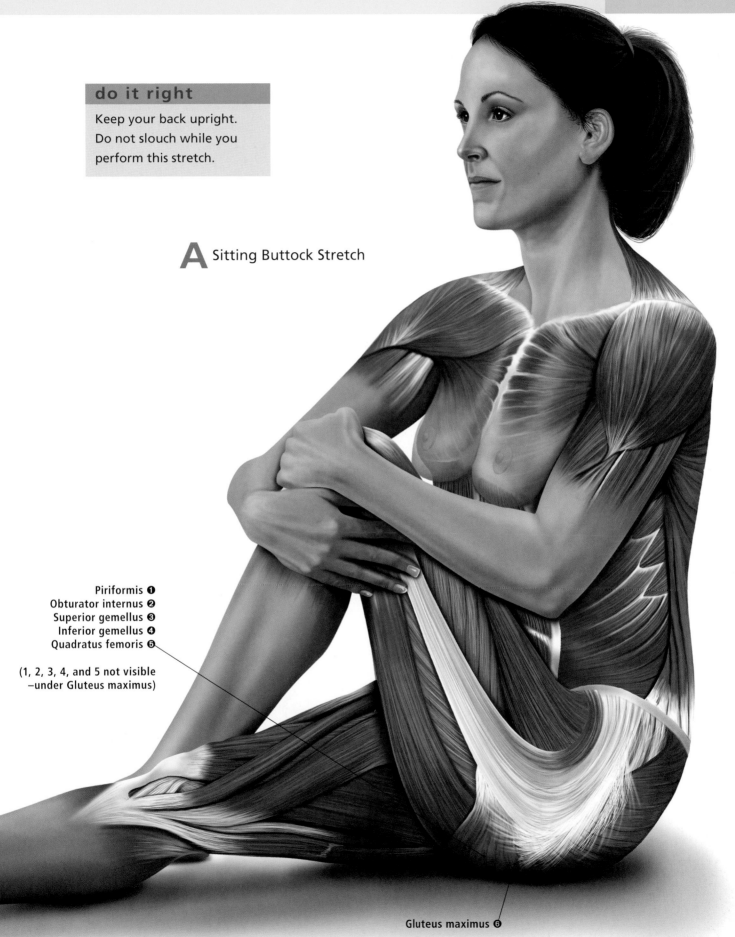

do it right
Keep your back upright.
Do not slouch while you
perform this stretch.

A Sitting Buttock Stretch

Piriformis ❶
Obturator internus ❷
Superior gemellus ❸
Inferior gemellus ❹
Quadratus femoris ❺

(1, 2, 3, 4, and 5 not visible
–under Gluteus maximus)

Gluteus maximus ❻

Standing Knee-to-chest and Pigeon Hip Stretches

The standing knee-to-chest stretch is probably more suited as a quick pull up or dynamic movement rather than a sustained static hold. If you are trying to hold this position for a prolonged time, make sure you have good balance and are not at any risk of falling. If you are at all unsteady on your feet, it is wise to have a companion help to steady you. Performing stretches in a more dynamic fashion can be very useful. They are less about improving range of motion (although they can help with this in some circumstances) and more about preparing the mind and body for upcoming activity. If this is the aim of your stretching, it is important to perform this type of movement as close as possible to the actual performance.

The pigeon stretch is a quite advanced stretch of the hip rotators and lateral thigh. It is also a popular yoga pose. It uses body weight to provide a quite intense stretch. The main target of the pigeon stretch is the piriformis muscle that lies deep in the gluteals. It is important in running sports and when performing quick changes of direction. Tightness at the hip can lead to poor alignment at the knee and many common knee and ankle injuries.

how to

The standing knee-to-chest stretch involves standing on one leg and bringing the other knee up toward the chest. Pull the knee and hip inwards and upwards, and feel the stretch in the gluteus maximus muscle.

The pigeon stretch starts in a push-up position, on hands and toes. Slide one knee up toward your hand, and rotate your hip so that your leg comes underneath your body and the outside of your ankle rests on the floor. Slide your other leg backward to move your hips toward the floor. Feel a deep stretch through the hip of the bent leg and in the lower back.

variations

ONE A The chair stretch is a good alternative for office workers to easily stretch their gluteals and piriformis. Simply sit on your chair with one leg crossed so that the ankle rests on the opposite knee. If needed, lean forward onto your desk to increase the stretch.

ONE B This extension of the pigeon stretch involves reaching your arms forward rather than staying in an upright position. To do this, bend forward, fully extend your arms in front of you, and rest your forearms on the floor.

muscles being stretched

❶ Iliotibial tract

❷ Gluteus maximus

❸ Piriformis (hidden under Gluteus maximus)

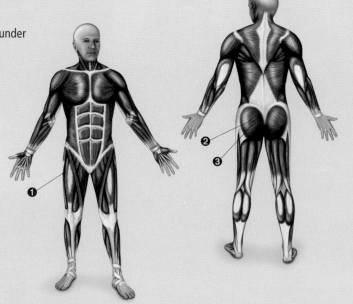

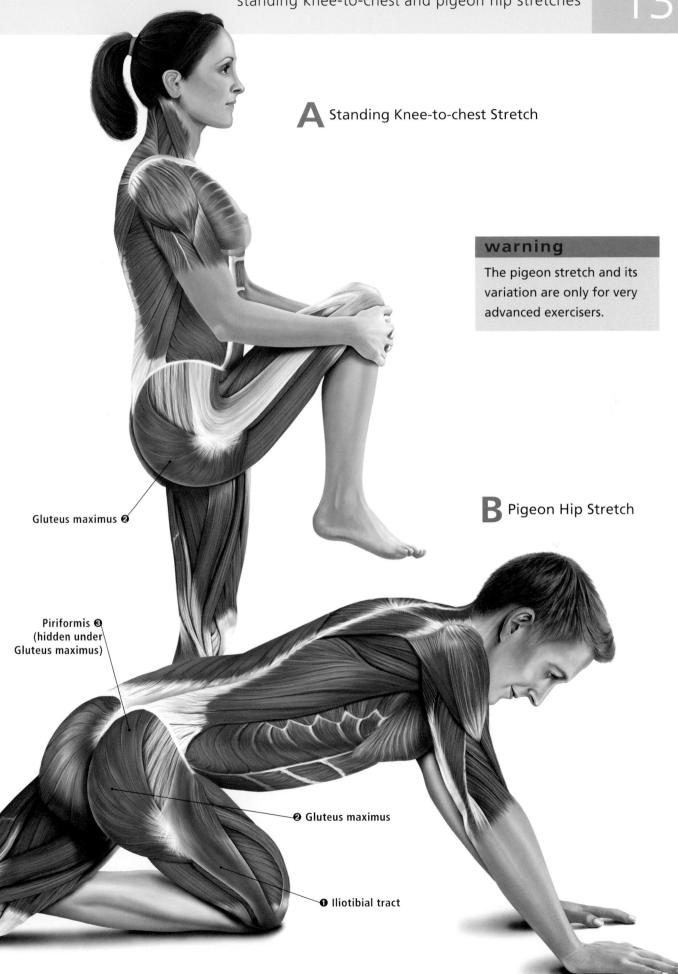

A Standing Knee-to-chest Stretch

warning

The pigeon stretch and its variation are only for very advanced exercisers.

B Pigeon Hip Stretch

Gluteus maximus ❷

Piriformis ❸
(hidden under
Gluteus maximus)

❷ Gluteus maximus

❶ Iliotibial tract

Standing Outer Hip and Standing Iliotibial Stretches

These stretches will not only stretch the gluteal muscles but also the small tensor fasciae latae muscle and its long tendon, the iliotibial tract or band (ITB). These stretches require no special equipment. ITB tightness is a common condition among dancers and long-distance runners, and is a leading cause of anterior knee pain. Following major knee injuries or surgery, tightness of the ITB can be an impediment to regaining full knee flexion. The sensation when stretching the ITB is different from other stretches, as the ITB is not muscle tissue but connective tissue with quite different elastic properties.

how to

To perform the standing outer hip stretch, stand with one leg bent at the hip and knee, and slightly bend the other knee as if performing a small single leg squat. Rotate the hip to turn the lifted leg and place the lateral side of the ankle onto the other knee. Bend slightly at the waist and bend the standing leg enough to feel the stretch in the hip.

The standing ITB stretch is performed by standing with one leg crossed behind the other. Lean with your shoulders away from the crossed-over leg and push your hips out to the other side. The leg placed behind is the leg that is being stretched.

variations

ONE A The standing outer hip stretch is quite challenging, and not always practical to perform. A very simple variation is to simply sit cross legged on the floor. This has the added advantage of stretching both hips at the same time.

ONE B Try performing a lying ITB stretch. To do this, lie down on the side that you want to stretch, keeping the bottom leg straight. Keep your back and hips straight, cross your top leg over the bottom leg, bend your knee, and lay it flat on the ground. Now position your bottom leg so that it is resting on a stool or chair. The great thing about this stretch is that you can lie there for as long as you like!

muscles being stretched

❶ Tensor fasciae latae
❷ Iliotibial tract
❸ Gluteus medius
❹ Gluteus minimus
❺ Gluteus maximus
❻ Biceps femoris

warning

Make sure you have good balance and a stable surface when performing any standing stretches. If you are at all unsteady, ask a companion to help.

do it right

When stretching the ITB, your body must stay in a straight line from foot to head when looking side on.

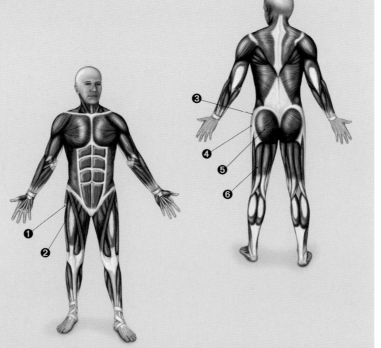

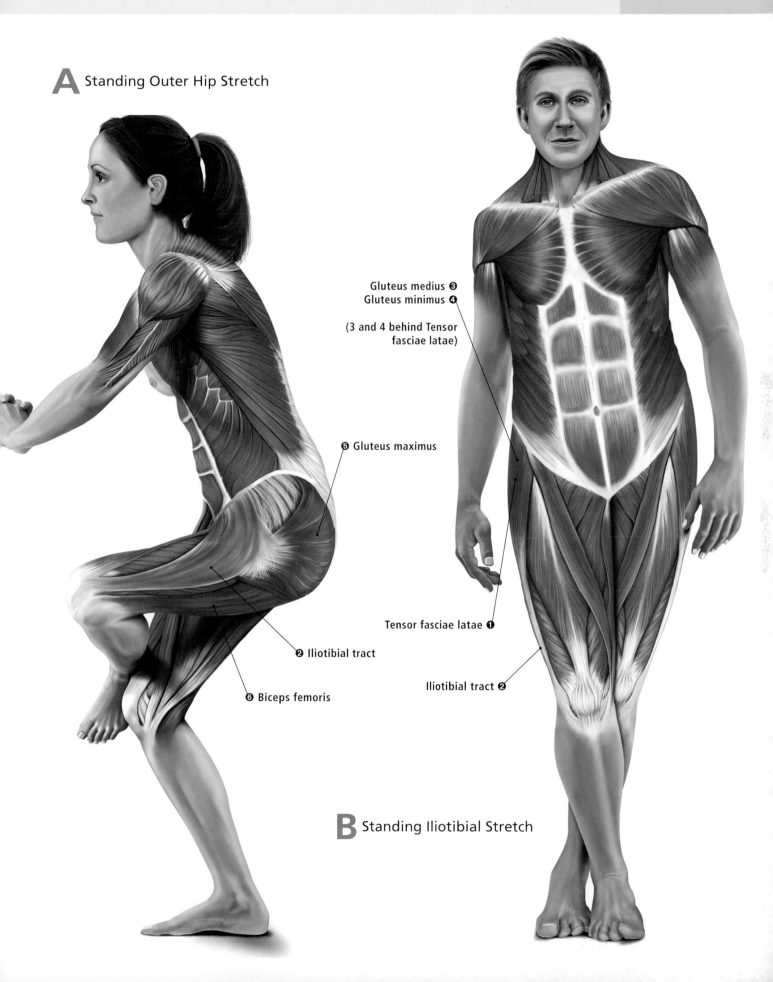

A Standing Outer Hip Stretch

Gluteus medius ❸
Gluteus minimus ❹

(3 and 4 behind Tensor
fasciae latae)

❺ Gluteus maximus

❷ Iliotibial tract

❻ Biceps femoris

Tensor fasciae latae ❶

Iliotibial tract ❷

B Standing Iliotibial Stretch

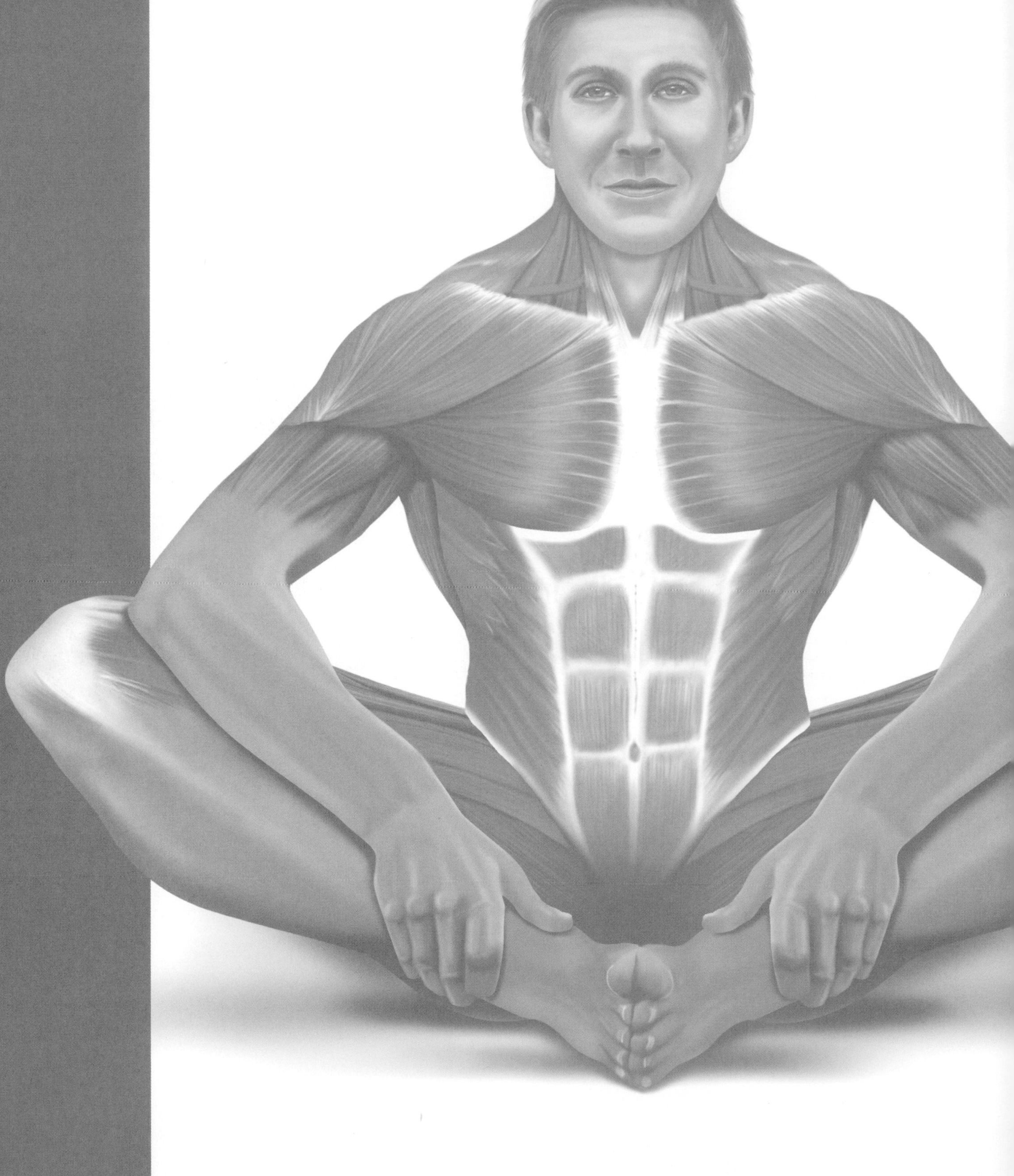

Thigh Stretches

The upper leg, or thigh, includes the quadriceps group in the anterior thigh, hamstrings in the posterior thigh, and adductors in the medial thigh. These are some of the most commonly strained muscle groups in athletes, especially those involved in sports that involve quick changes of pace or changes of direction. Appropriate flexibility is needed for kicking athletes and for gymnasts, dancers, and divers. The quadriceps group includes vastus medialis, vastus lateralis, vastus intermedius, and rectus femoris. The hamstrings include biceps femoris, semimembranosus, and semitendinosus. The adductor group consists of adductor magnus, adductor longus, and adductor brevis. Smaller muscles in the thigh include gracilis and sartorious.

Joints and Muscles of the Thigh and Leg

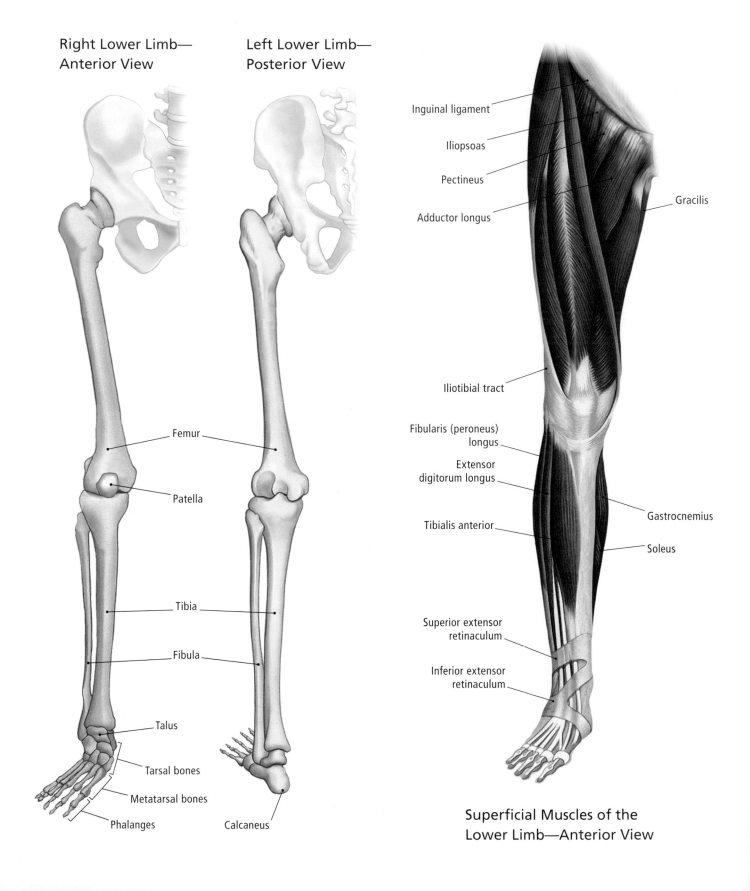

Right Lower Limb—
Anterior View

Left Lower Limb—
Posterior View

Inguinal ligament

Iliopsoas

Pectineus

Adductor longus

Gracilis

Iliotibial tract

Fibularis (peroneus)
longus

Extensor
digitorum longus

Gastrocnemius

Tibialis anterior

Soleus

Superior extensor
retinaculum

Inferior extensor
retinaculum

Femur

Patella

Tibia

Fibula

Talus

Tarsal bones

Metatarsal bones

Phalanges

Calcaneus

Superficial Muscles of the
Lower Limb—Anterior View

Muscles of the Thigh—Transverse Section

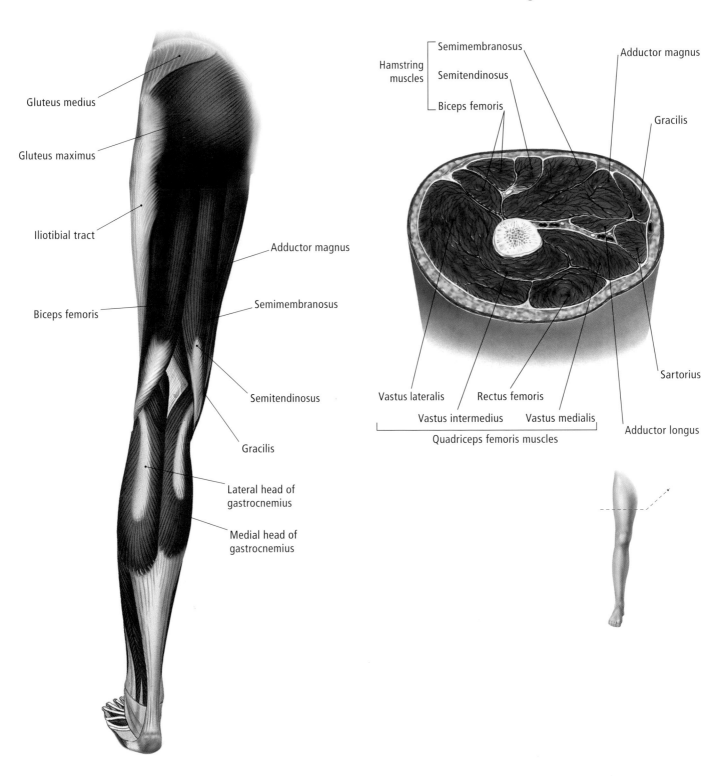

Gluteus medius

Gluteus maximus

Iliotibial tract

Biceps femoris

Adductor magnus

Semimembranosus

Semitendinosus

Gracilis

Lateral head of
gastrocnemius

Medial head of
gastrocnemius

Hamstring
muscles

Semimembranosus

Semitendinosus

Biceps femoris

Adductor magnus

Gracilis

Sartorius

Adductor longus

Vastus lateralis

Vastus intermedius

Rectus femoris

Vastus medialis

Quadriceps femoris muscles

**Superficial Muscles of the Lower
Limb—Posterior View**

Standing Quadriceps Stretch

This classic quadriceps stretch has been a favorite for generations. It is easy to perform and can be done incidentally while out walking or waiting for the bus. The main muscles stretched in this position are vastus medialis, vastus lateralis, and vastus intermedius. Rectus femoris and iliopsoas will also receive a good stretch if the hip is held in extension. If you hold the foot rather than the ankle, you will also feel a stretch through tibialis anterior along the front of your shin. Tight quadriceps will limit knee flexion and hip extension, and may pull the pelvis into an anterior rotated position that will increase the curve in the lower back.

warning

Try one of the variations if you experience any back or knee pain during the primary stretch. If your balance is poor, get help from an assistant or try a variation.

how to

Stand in a balanced level position. Bend one knee up and hold your foot or ankle with either hand (whichever feels more comfortable), pulling your heel in toward your buttock. (If you are at all unsteady on your feet, make sure you are close to something secure you can grab with your other hand.) As you pull your foot back, you should feel a comfortable stretch through the front of your thigh.

variations

ONE

A If you experience back pain doing this stretch, or have trouble balancing, then try the same stretch lying down on your side. Lie on the opposite side to the leg you wish to stretch. Then perform the same movement as above: bend the knee and grasp the foot or ankle while pulling in toward the buttock.

TWO

A If your knee doesn't bend far enough to be able to reach with your hand, then try this variation. Stand in the same position as above, but rather then holding your foot or ankle, place the foot onto a chair, bench, or something similar behind you. Find an object you can rest your foot on that is at a comfortable position, but enough to feel a stretch. For a gentle increase in the stretch, bend your standing knee slightly.

muscles being stretched

❶ Rectus femoris

❷ Vastus lateralis

❸ Vastus intermedius
   (deep to Rectus femoris)

❹ Vastus medialis

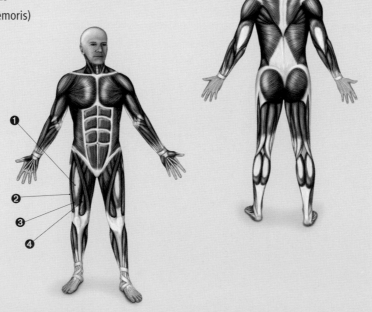

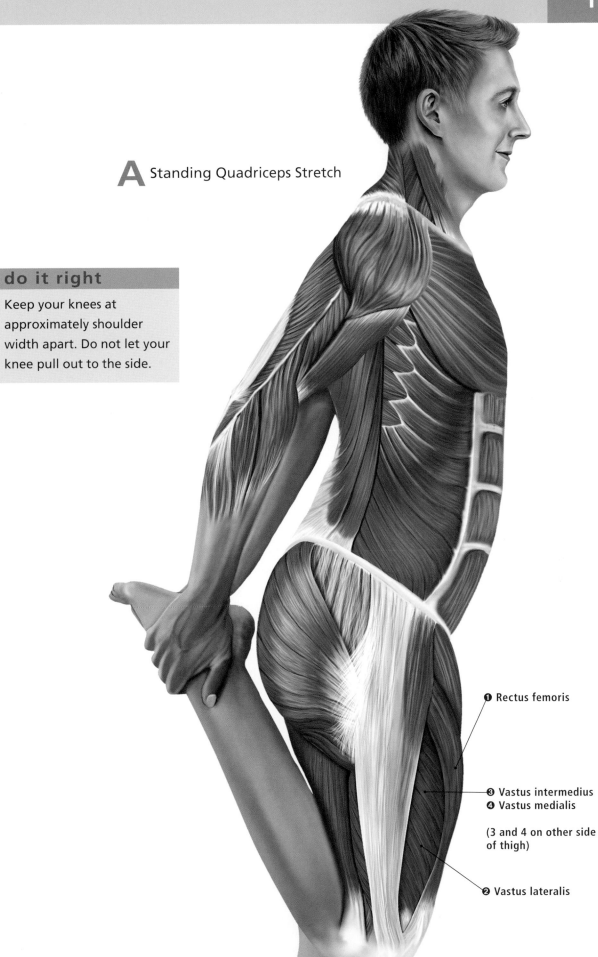

A Standing Quadriceps Stretch

do it right

Keep your knees at approximately shoulder width apart. Do not let your knee pull out to the side.

❶ Rectus femoris

❸ Vastus intermedius
❹ Vastus medialis

(3 and 4 on other side of thigh)

❷ Vastus lateralis

Sitting One-leg and Standing Toe-up Hamstring Stretches

The hamstrings seem to be one of the muscles or muscle groups in the human body that are most pone to tightness and many people enjoy giving them a good stretch. One of the most common stretches is the sitting one-leg stretch. The standing toe-up stretch is great, as it can be done as an incidental stretch during the day without much thought or preparation. The main target areas for these stretches are the hamstrings, but other muscles that will be stretched include sartorius, gastrocnemius, and flexor digitorum longus. Anyone who experiences sciatic pain should be very careful with these stretches and consult expert opinion before trying them. It is possible to modify these stretches to effectively stretch the muscle tissue but not aggravate the nerve tissue.

how to

The sitting one-leg stretch is performed while seated. Starting in a long-sit position, keep the leg you wish to stretch straight and bend the other leg so that the sole of the foot slides up against the knee of the other leg. Bend forward at the waist and reach toward your toes.

The standing toe-up stretch is best performed with your toes resting up against a wall or step, but this is not essential. The stretched leg should be pointed out in front with the toes and ankle pulled back toward the knee. Bend the other knee and lean forward at the waist, feeling the stretch behind the knee.

variations

ONE A The sitting one-legged stretch can be performed by pulling on a towel, or stretch band, over the toes. This will increase the intensity of the stretch and will also include gastrocnemius into the muscles being stretched.

ONE B Try standing with your leg raised up onto a bench or fence rail. This variation combines the two positions by being a standing stretch but with the hamstrings being stretched at hip height. To do this, place one heel up on the fence or back of the bench, slightly bend the other knee, if needed, then bend forward from the waist.

muscles being stretched

❶ Biceps femoris
❷ Semimembranosus
❸ Semitendinosus

do it right

Bend forward from the waist rather than arching the upper back and neck.

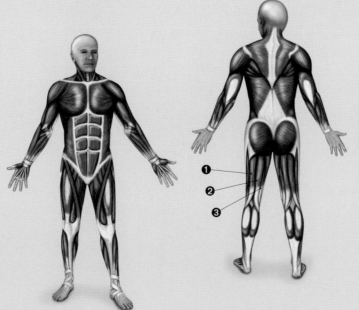

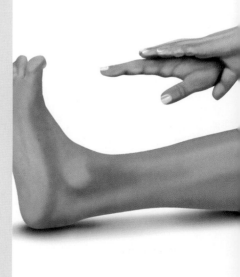

B Standing Toe-up Hamstring Stretch

warning
Bend the knee slightly if there is any discomfort when performing this stretch. Seek medical advice if you experience pain while doing this stretch.

Biceps femoris ❶

Semimembranosus ❷
Semitendinosus ❸

(2 and 3 hidden on inner side of thigh)

A Sitting One-leg Hamstring Stretch

Biceps femoris ❶
(hidden on other side of thigh)

Semitendinosus ❸

❷ Semimembranosus

Lying Hamstring Stretch

The lying hamstring stretch is a great one for people who have limited hamstring flexibility. The reason for this is that, in the lying position, there is no pressure pushing down on the knee, so the stretch can be completed with even very limited knee extension. The other great thing about this position is that the leg can rest on a support in the stretch position, enabling a very prolonged stretch, if so desired. This stretch can also be performed as a partner-assisted stretch, a neural glide technique, or a PNF (proprio-neuromuscular-facilitation) technique. The three hamstring muscles will be stretched equally in this position.

warning

If your head does not comfortably lie on the ground, use a towel or pillow underneath your head as support. Seek medical advice if you experience pain while doing this stretch.

how to

Start by lying on your back. Bend the hip and knee of one leg, keeping the other leg straight. Use your hands to support the leg behind the knee. From this position, straighten your leg as far as you are able, without changing your hip position. You should feel the stretch behind your knee.

variations

ONE

A Try a partner-assisted stretch. The person stretching lies on the ground and lifts his or her leg as above but rests it on the shoulder of the partner. The partner can hold the stretched leg at the ankle and knee, and move forward far enough so that the stretch is felt. To progress this to a PNF stretch, the stretcher lightly pushes down into the shoulder of his or her partner for a few seconds and then relaxes. During the relax phase, the partner pushes the leg into further stretch. Repeat as many times as desired.

TWO

A For a neural glide (nerve stretch) rather than a muscle stretch, adopt the same position; but rather than straightening the leg as far as you can, straighten just short of this point and then bring your leg down to a relaxed position. Keep moving your leg back and forth at the knee to "glide" the nerve.

muscles being stretched

❶ Gluteus maximus

❷ Biceps femoris

❸ Semimembranosus

❹ Semitendinosus

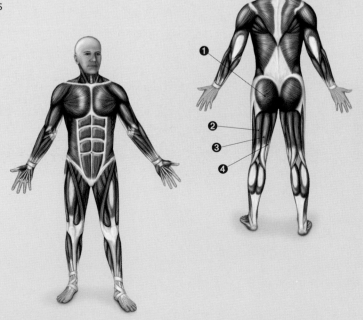

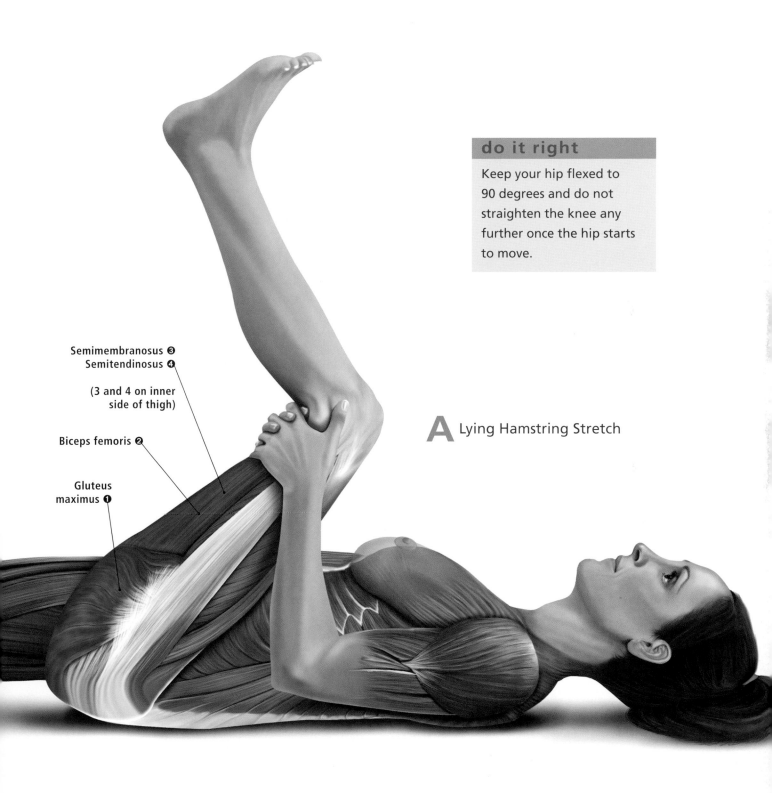

do it right

Keep your hip flexed to 90 degrees and do not straighten the knee any further once the hip starts to move.

Semimembranosus ❸
Semitendinosus ❹

(3 and 4 on inner side of thigh)

Biceps femoris ❷

Gluteus maximus ❶

A Lying Hamstring Stretch

Standing Hamstring and Sitting Adductor Stretches

The standing hamstring stretch is a good test of the effect of your hamstring length on your daily activities. If you can only reach to your knees or just below, your hamstrings are very tight and likely pulling your pelvis into a posteriorly rotated position. This can be a contributor to lower back pain and poor posture. The lower back muscles will also be stretched as you reach forward.

The sitting adductor stretch has the advantage of stretching the left and right adductor group at the same time. This position will also enable you to see if there is an imbalance between the left and right side. An imbalance of the adductors could lead to shifting of the pubic symphysis and consequently problems such as osteitis pubis.

warning

Bend your knees slightly when performing the standing hamstring stretch if you feel any discomfort in your back or thigh while stretching. Seek medical advice if you experience pain.

how to

For the standing hamstring stretch, start by standing with your feet pointing forward, shoulder-width apart. Lean forward and run your hands down the front of your thighs. At the knees, your hands will come away from your legs. Keep leaning until you feel the stretch in the back of your legs. If there is any discomfort, you can bend your knees slightly.

For the sitting adductor stretch, start by sitting on the floor. Bend your hips and knees, and bring the soles of your feet together. Hold your feet and rest your elbows on the inside of your knees. To increase the stretch, you can push down on your knees with each elbow.

variations

ONE

B Change the sitting adductor stretch to a lying stretch. Your lower body will be in exactly the same position but instead of sitting upright, lie back and relax! The great thing about this stretch is the use of gravity to gradually increase the intensity of the stretch.

TWO

B The "V" sit stretch will also stretch the left and right adductors together. This time, sit with your legs straight out in front of you. Now move each leg out to the side into a "V" position, feeling the stretch equally on left and right.

do it right

Keep your back straight and your chin away from your chest while performing these stretches.

muscles being stretched

❶ Pectineus

❷ Aductor longus

❸ Adductor brevis

❹ Adductor magnus

❺ Gracilis

❻ Gluteus maximus

❼ Biceps femoris

❽ Semimembranosus

❾ Semitendinosus

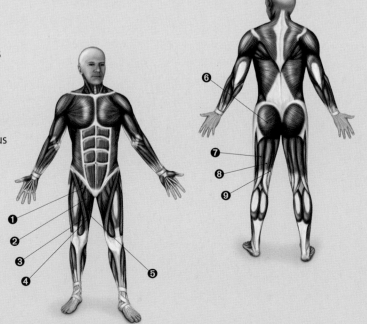

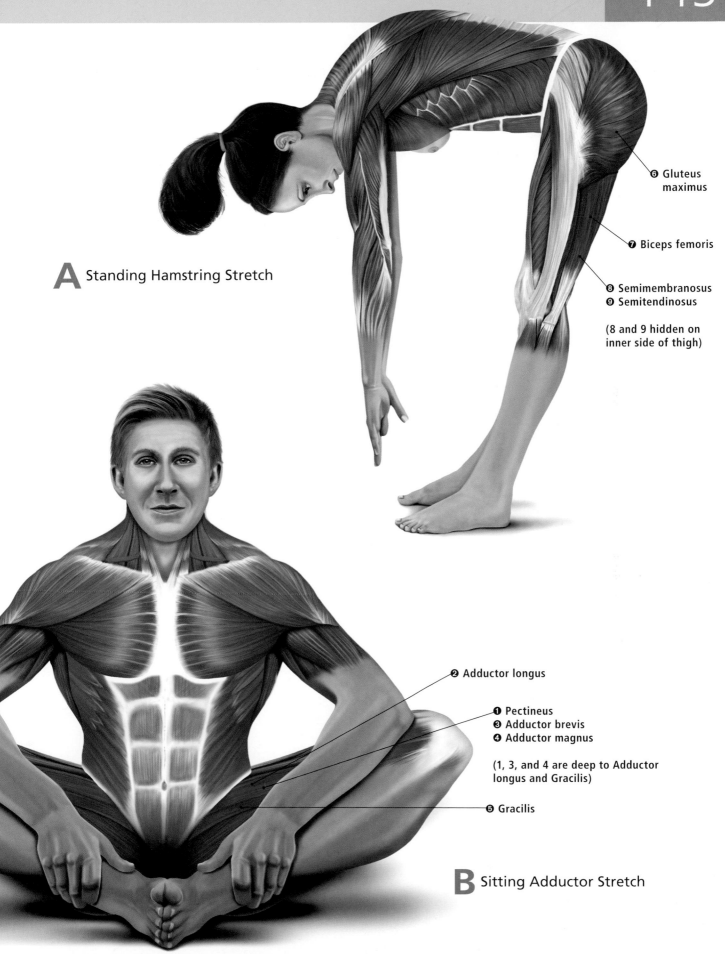

A Standing Hamstring Stretch

❻ Gluteus maximus

❼ Biceps femoris

❽ Semimembranosus
❾ Semitendinosus

(8 and 9 hidden on inner side of thigh)

❷ Adductor longus

❶ Pectineus
❸ Adductor brevis
❹ Adductor magnus

(1, 3, and 4 are deep to Adductor longus and Gracilis)

❺ Gracilis

B Sitting Adductor Stretch

Side Lunge Adductor Stretch

This stretch is for the adductor muscles, as well as gracilis. These muscles are important for many sports, especially those that include kicking and quick changes of direction while running. Dancers and gymnasts need the flexibility to be able to do "the splits." People who spend a lot of time sitting will typically find that these muscles are tight. This stretch can be performed as a static stretch for flexibility, or dynamically as part of a sports warm-up. Tight adductor muscles limit the ability to abduct the leg (meaning to move it sideways away from the body) and can lead to inward rotation of the knee, increasing the risk of knee and hip injury.

warning

Keep your knee bent directly over your middle toes to avoid the risk of knee pain.

how to

Start by standing with hands on hips and feet greater than shoulder-width apart, toes pointing forward. Bend one knee and shift your body weight over the bent knee. Feel the stretch along the inside of the opposite leg. If you do not feel the stretch, start with your feet further apart.

variations

ONE

A Try a standing leg-up adductor stretch. Stand side on to a bench. Lift one leg up and out to the side, resting the inside of your foot on the back of the bench. For most people, this will be enough of a stretch; but, if not, bend the other knee and shift your weight as above.

TWO

A Try a standing adductor stretch with a stability ball for extra challenge. Stand with one leg bent and your knee resting on the stability ball. Slowly roll the ball away from you until you feel the stretch in your inner thigh. Take care and make sure you can keep the ball in control: this is an advanced exercise.

muscles being stretched

❶ Pectineus

❷ Adductor longus

❸ Adductor brevis

❹ Adductor magnus

❺ Gracilis

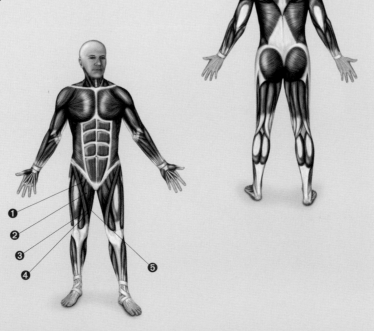

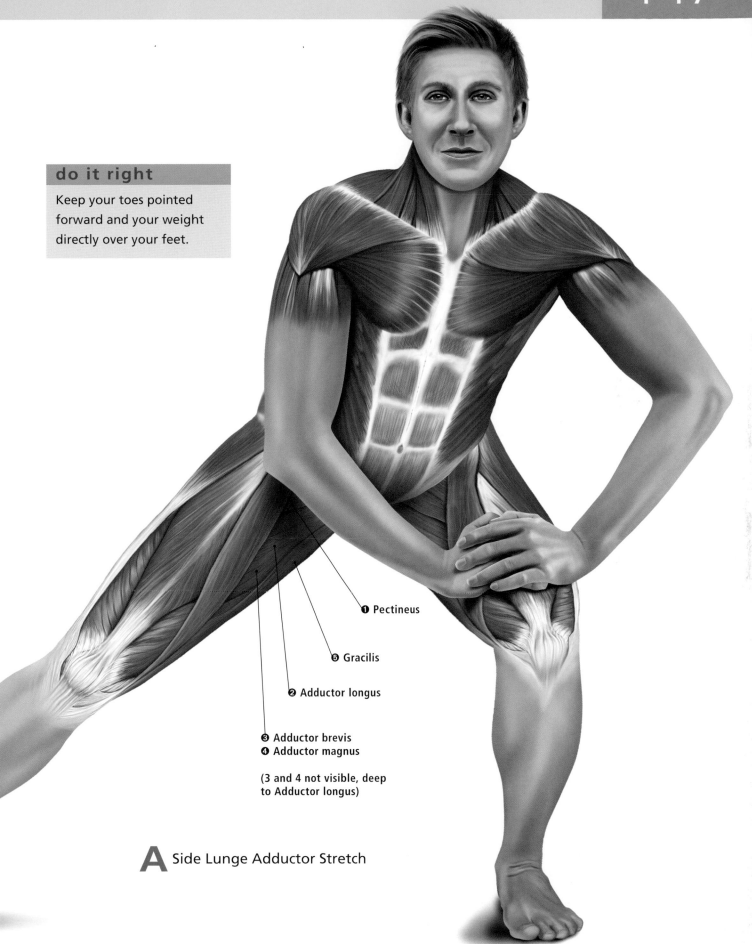

do it right

Keep your toes pointed forward and your weight directly over your feet.

❶ Pectineus

❺ Gracilis

❷ Adductor longus

❸ Adductor brevis
❹ Adductor magnus

(3 and 4 not visible, deep to Adductor longus)

A Side Lunge Adductor Stretch

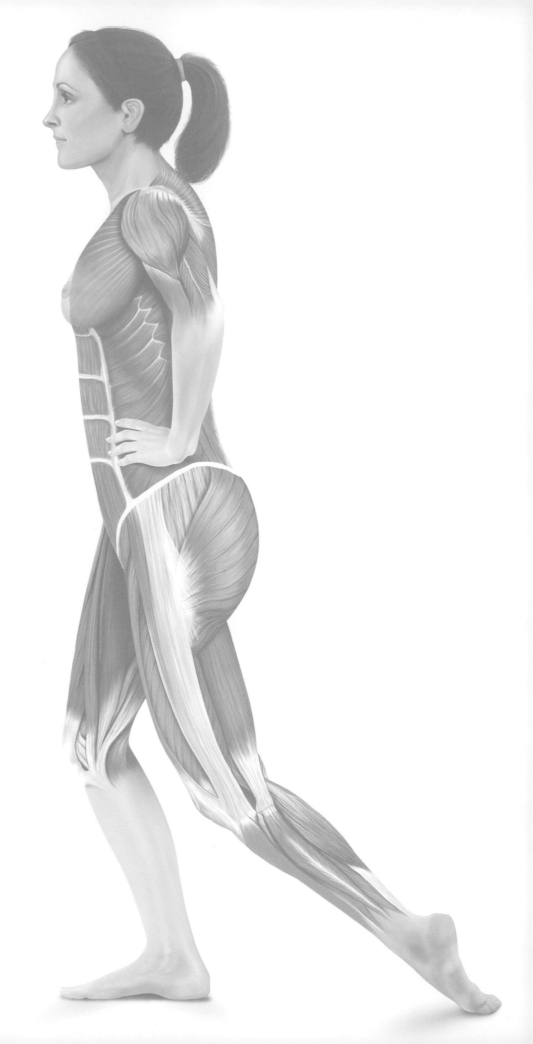

Leg and Foot Stretches

Muscles in the lower leg and foot include the large calf muscles, gastrocnemius and soleus, and the smaller popliteus and plantaris. Tibialis posterior, the peroneal (fibularis) muscles, and toe flexors lie postero-laterally. Tibialis anterior and the toe extensors are at the front of the leg. In the foot are the intrinsic muscles that support the arch, and small muscles that help move the toes. As the foot and ankle are our connection to the ground, it is vitally important to keep these muscles healthy. Problems with the position of the feet can lead to foot pain, knee pain, hip pain, and lower back pain, and could contribute to many other common conditions.

Joints and Muscles of the Leg, Foot, and Ankle

Deep Muscles of the Lower Limb— Anterior View

Deep Muscles of the Lower Limb— Posterior View

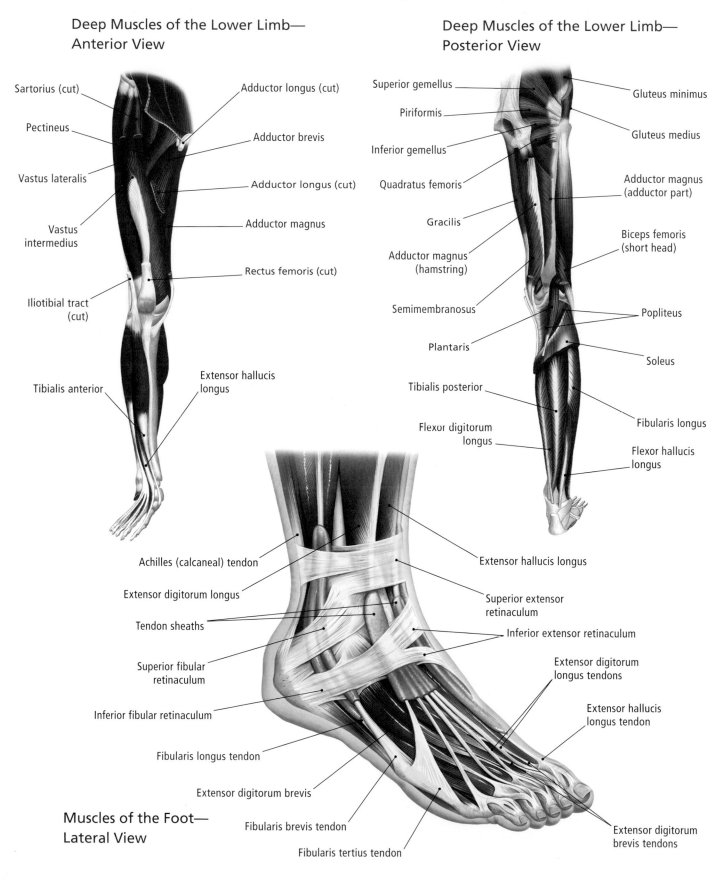

Sartorius (cut)

Pectineus

Vastus lateralis

Vastus intermedius

Iliotibial tract (cut)

Tibialis anterior

Adductor longus (cut)

Adductor brevis

Adductor longus (cut)

Adductor magnus

Rectus femoris (cut)

Extensor hallucis longus

Superior gemellus

Piriformis

Inferior gemellus

Quadratus femoris

Gracilis

Adductor magnus (hamstring)

Semimembranosus

Plantaris

Tibialis posterior

Flexor digitorum longus

Gluteus minimus

Gluteus medius

Adductor magnus (adductor part)

Biceps femoris (short head)

Popliteus

Soleus

Fibularis longus

Flexor hallucis longus

Achilles (calcaneal) tendon

Extensor digitorum longus

Tendon sheaths

Superior fibular retinaculum

Inferior fibular retinaculum

Fibularis longus tendon

Extensor digitorum brevis

Muscles of the Foot— Lateral View

Fibularis brevis tendon

Fibularis tertius tendon

Extensor hallucis longus

Superior extensor retinaculum

Inferior extensor retinaculum

Extensor digitorum longus tendons

Extensor hallucis longus tendon

Extensor digitorum brevis tendons

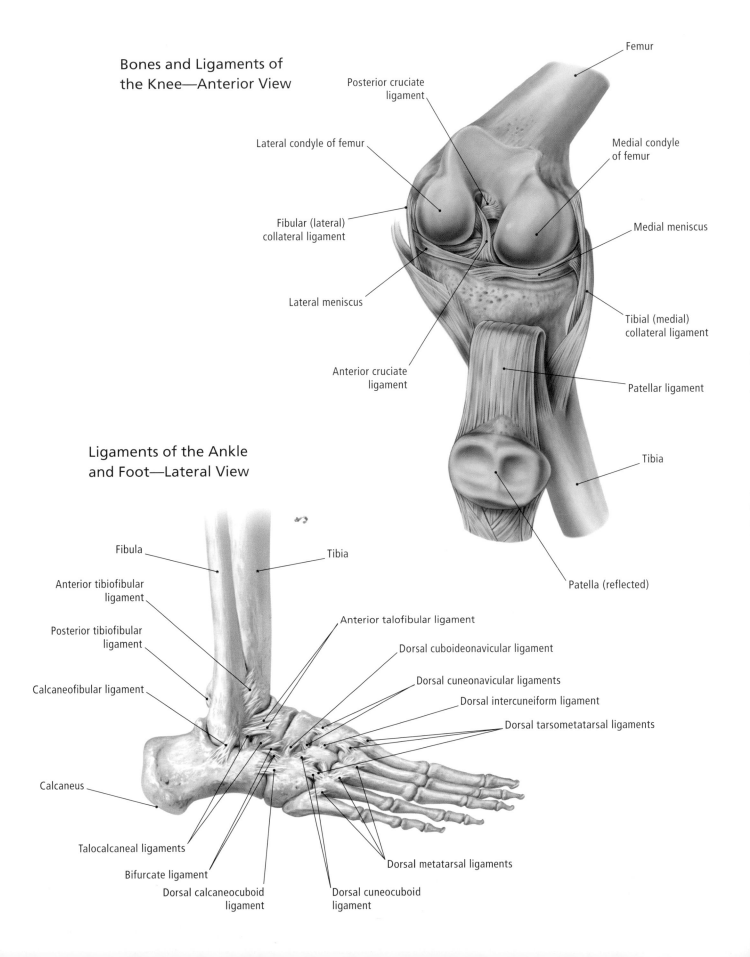

Bones and Ligaments of the Knee—Anterior View

Femur

Posterior cruciate ligament

Lateral condyle of femur

Medial condyle of femur

Fibular (lateral) collateral ligament

Medial meniscus

Lateral meniscus

Tibial (medial) collateral ligament

Anterior cruciate ligament

Patellar ligament

Tibia

Patella (reflected)

Ligaments of the Ankle and Foot—Lateral View

Fibula

Tibia

Anterior tibiofibular ligament

Posterior tibiofibular ligament

Anterior talofibular ligament

Dorsal cuboideonavicular ligament

Dorsal cuneonavicular ligaments

Dorsal intercuneiform ligament

Calcaneofibular ligament

Dorsal tarsometatarsal ligaments

Calcaneus

Talocalcaneal ligaments

Bifurcate ligament

Dorsal metatarsal ligaments

Dorsal calcaneocuboid ligament

Dorsal cuneocuboid ligament

Heel Drop Stretch

The heel drop stretch utilizes a step to stretch the gastrocnemius either singularly or both together. Soleus, popliteus, and the Achilles tendon are other areas that will be stretched. This stretch will help improve ankle dorsi-flexion, which is important when walking on hills or stairs. Poor dorsi-flexion can cause anterior ankle pain and is often reduced following injury. After periods of immobilization, restoring dorsi-flexion is a major goal of recovery. While there is no evidence that static stretches of the gastrocnemius will improve sports performance, dynamic or ballistic stretching has been shown to have a beneficial effect on vertical jump height in basketball players.

warning

Static calf stretching can actually be detrimental to sports performance. Talk to your sports professional to ensure that this stretch is right for you.

how to

Start by standing on a step facing upwards. Move one heel off the back of the step so that the ball of your foot rests on the step. Keep your body upright and your knee straight. Bend your other knee and drop your heel to below the level of the step. Feel the stretch through your calf.

variations

ONE

A A stretch band is a great piece of equipment to enhance your stretching. To stretch your calf with a stretch band, step through the loop of the band and pull the band up to your waist. Lie down on your back and place one leg inside the loop of the band, and place the band over your toes. Straighten out your leg, then lift it to the angle where you feel a good stretch through the calf and hamstring.

TWO

A A double heel drop stretch will maximize your time by stretching both calves together. To perform this, stand as for the single heel drop stretch, but on the balls of both feet. Have both heels off the step and bend at the ankles to drop the heels off the edge of the step. Keep both knees straight. Take care if you have balance problems. You may need to hold onto something to steady you.

muscles being stretched

❶ Gastrocnemius

❷ Soleus

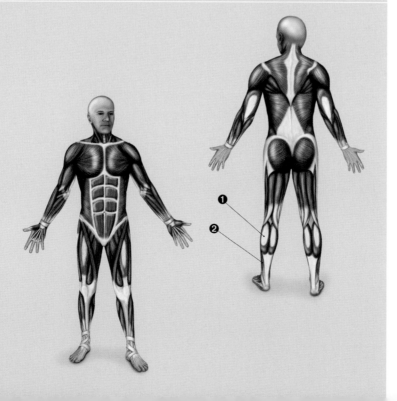

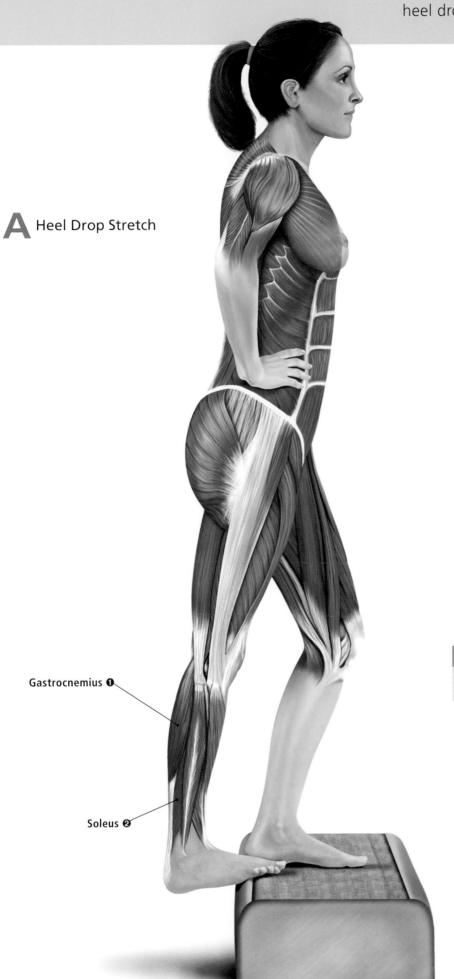

A Heel Drop Stretch

Gastrocnemius ❶

Soleus ❷

Standing Shin and Standing Achilles Stretches

The main target muscles of these stretches are the tibialis anterior and soleus muscles. The tibialis anterior is the largest muscle in the shin and is a dorsi-flexor of the foot. The soleus muscle lies deep under the gastrocnemius and is a plantar-flexor of the foot. The soleus attaches to the heel via the large Achilles tendon. Both of these stretches are performed while standing and could easily be performed one after the other. They are well suited to pre- and post-exercise routines.

warning

A painful Achilles tendon is unlikely to respond well to vigorous stretching.

how to

For the standing shin stretch, start by standing in a wide stance with one foot forward of the other, toes pointing forward. Bend your front knee, shift your weight onto the front leg, point your other foot back toward the ground, and place the toes on the ground. Shift your weight further forward until you feel the stretch in the front of your shin.

For the Achilles stretch, start in the same stance as for the standing shin stretch, but this time keep the soles of both feet flat on the ground. This time, shift your weight onto the back leg, bend the knee, and push the heel down into the ground. You should feel the stretch in the lower part of the calf.

variations

ONE A Try a kneeling shin stretch, which is ideal for performing at home, on a carpeted floor, blanket, or mat. Point your toes back behind you and sit back gently onto your heels. Be careful if you have any preexisting knee pain.

ONE B A strong, intense stretch of the calf and Achilles can be performed with a towel or stretch band. While seated with the leg straight out in front of you, place the towel around your toes. Grasp the towel at both ends and pull up until you feel the stretch through your foot and posterior leg.

muscles being stretched

❶ Extensor hallucis longus

❷ Extensor digitorum longus

❸ Peroneus tertius

❹ Tibialis anterior

❺ Gastrocnemius

❻ Soleus

do it right

Keep your heel down on the ground when performing the standing Achilles stretch.

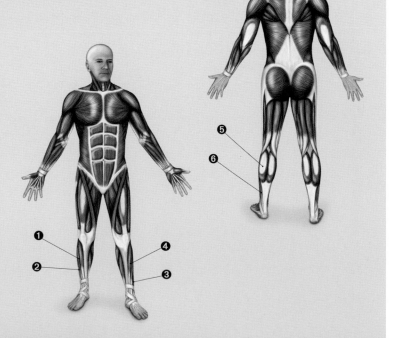

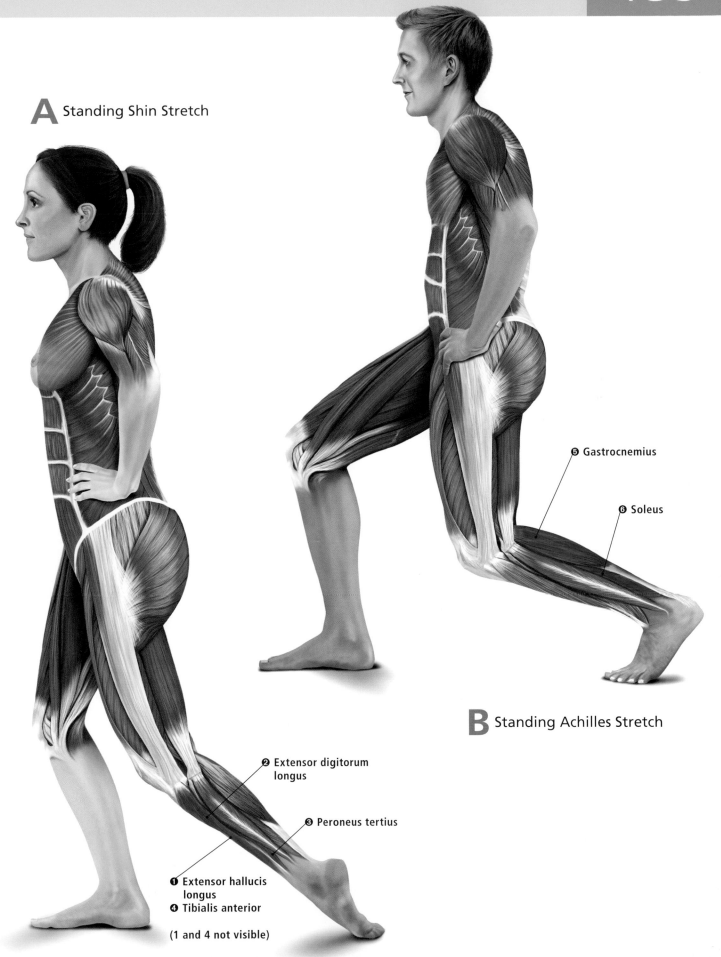

A Standing Shin Stretch

B Standing Achilles Stretch

❺ Gastrocnemius

❻ Soleus

❷ Extensor digitorum longus

❸ Peroneus tertius

❶ Extensor hallucis longus
❹ Tibialis anterior

(1 and 4 not visible)

Leaning Calf Stretch

The leaning calf stretch, or straight leg calf stretch, aims to stretch the gastrocnemius, the largest muscle in the calf. It is an easy stretch to perform on the go—you just need a wall or fence to lean against. While the main role of the gastrocnemius is as a plantar-flexor of the foot, it is also is involved in knee flexion. Therefore, for an effective stretch, you need to keep the knee in extension. The gastrocnemius forcefully contracts during running and jumping movements. Calf tightness can affect walking, especially up and down stairs, and can lead to ankle or shin pain.

warning

Calf pain or tightness could be a sign of a muscle tear. Vigorous stretching of an acutely torn muscle will make it worse. See your health professional if you experience any pain in the calf.

how to

Stand at arms length from the wall and lean forward, placing both hands on the wall approximately shoulder-width apart. Stand with one foot closer to the wall and the other back behind you. Keeping your back knee straight, lean into the wall from the hips until you feel a stretch through the calf along the back leg.

variations

ONE

A If you do not have a wall to lean against, you can just use the ground. Start from a push-up position, one foot crossed over the other and carry your weight through the ball of your foot. Push your heel back toward the ground. You should feel quite an intense stretch in this position.

TWO

A This stretch can be performed as a partner-assisted stretch. One person lies down on his or her back with one leg up in the air. The partner assists by pushing down on the toes, and can help maintain a straight knee at the same time. Make sure to communicate with your partner so that an appropriate stretch intensity is maintained.

muscles being stretched

❶ Gastrocnemius

❷ Soleus

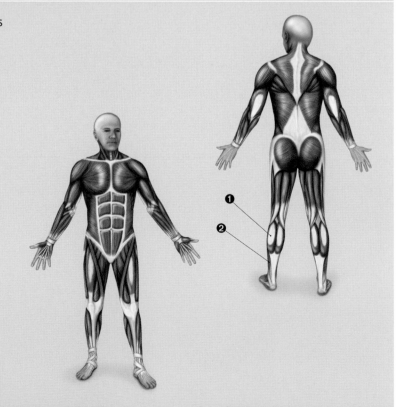

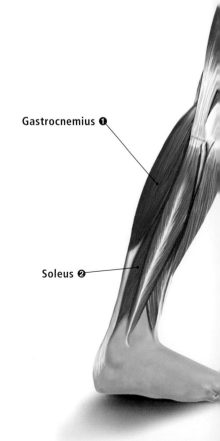

Gastrocnemius ❶

Soleus ❷

A Leaning Calf Stretch

do it right
Keep your knee straight and heel flat on the leg being stretched.

Ankle Rotation and Toe Squat Stretches

The ankle rotation exercise is more of a mobilization technique than a true stretch. No one muscle is specifically targeted; rather, the overall mobility and flexibility of the ankle joint is affected. This is an especially useful rehabilitation exercise following a period of immobilization—for example, after cast removal postfracture. The toe squat stretch is a stretch of the toe and foot plantar-flexors. It will stretch the small intrinsic muscles of the foot as well as flexor digitorum longus and flexor hallucis longus. This stretch is quite intense for the toe joints, so be careful not to stretch into pain.

warning

Ensure you have the strength and mobility to stand up again if you are performing the toe squat stretch. Do not stay in the position if you are feeling pain.

how to

Perform the ankle rotation while sitting or standing by simply rolling your foot and ankle around in a circle. First turn in one direction, then reverse to the other direction. This combines all the movements of the ankle in one motion.

For the toe squat stretch, start by kneeling on the ground. Flex your ankles to bring your feet to a position where you can curl the toes up. Try to rest on the balls of your feet rather than the toes. Sit back and rest on your heels.

variations

ONE A If your ankle is very stiff, then a manual ankle rotation stretch might be beneficial. This stretch is the same as a regular ankle rotation, except that you assist the movement with your hands. This stretch is safer to perform in a sitting position rather than standing.

ONE B If the toe squat stretch is too intense, try this variation. Instead of sitting back on your heels, perform the stretch in a crawling position. The feet will still be in the same position, but there will be no weight pushing down on them, and there is less need for knee and quadriceps flexibility.

do it right

Rotate the ankles clockwise and counterclockwise to ensure you get an optimal stretch.

muscles being stretched

❶ Extensor hallucis longus

❷ Extensor digitorum longus

❸ Peroneus longus

❹ Peroneus brevis

❺ Peroneus tertius

❻ Tibialis anterior

❼ Gastrocnemius

❽ Soleus

❾ Tibialis posterior

❿ Flexor digitorum longus

⓫ Flexor hallucis longus

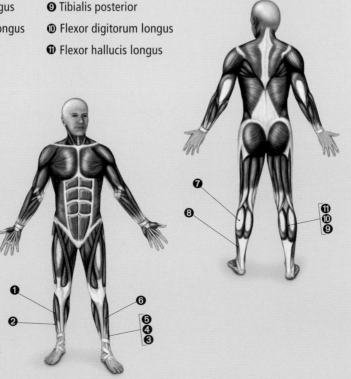

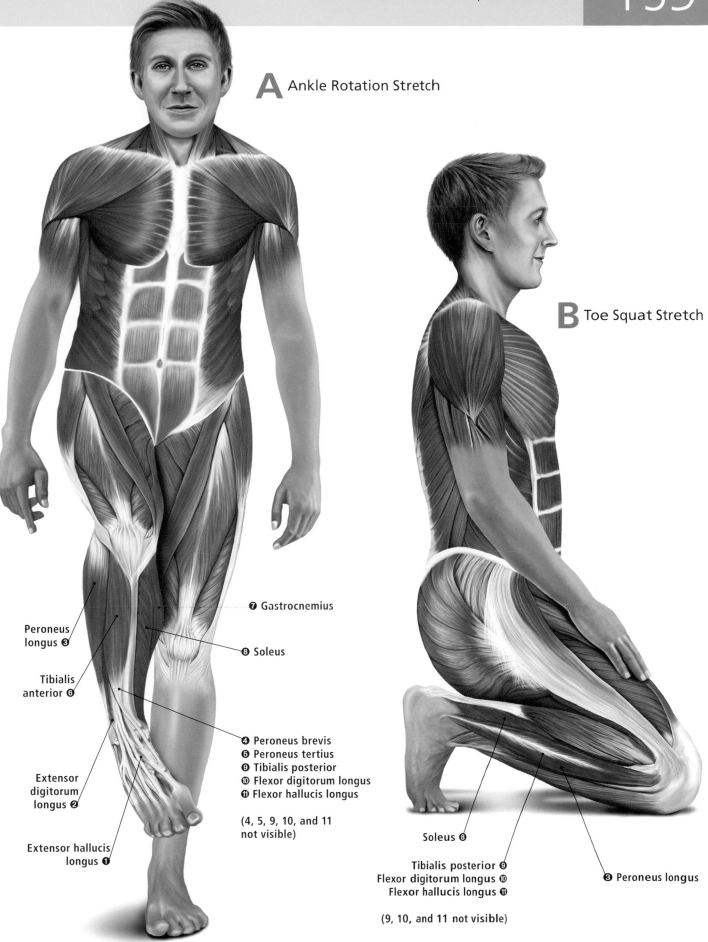

A Ankle Rotation Stretch

B Toe Squat Stretch

❼ Gastrocnemius

Peroneus
longus ❸

❽ Soleus

Tibialis
anterior ❻

❹ Peroneus brevis
❺ Peroneus tertius
❾ Tibialis posterior
❿ Flexor digitorum longus
⓫ Flexor hallucis longus

(4, 5, 9, 10, and 11
not visible)

Extensor
digitorum
longus ❷

Extensor hallucis
longus ❶

Soleus ❽

Tibialis posterior ❾
Flexor digitorum longus ❿
Flexor hallucis longus ⓫

(9, 10, and 11 not visible)

❸ Peroneus longus

Coloring workbook

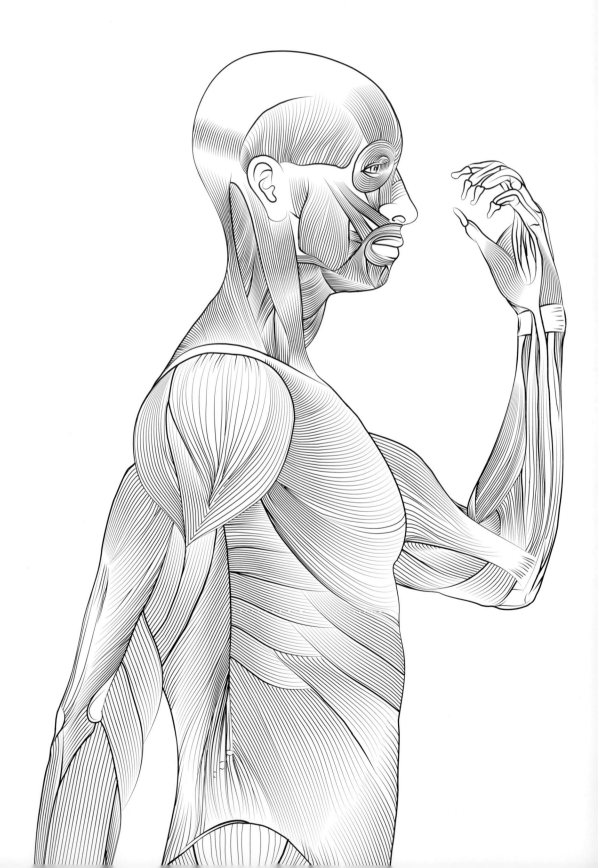

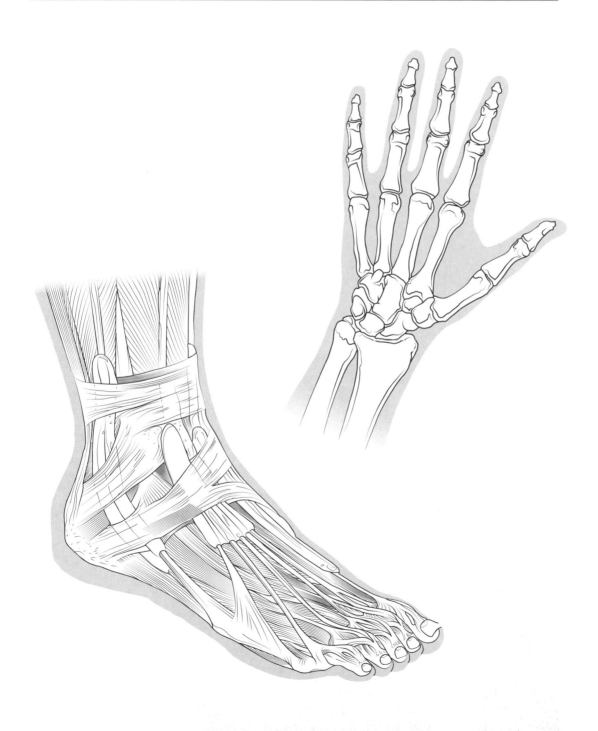

Muscular System

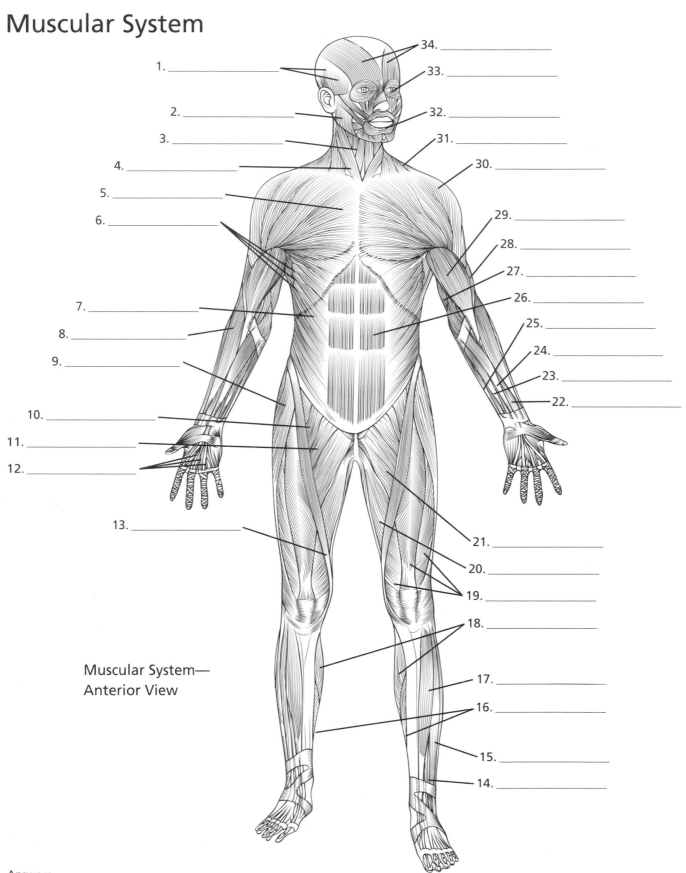

1. _____
2. _____
3. _____
4. _____
5. _____
6. _____
7. _____
8. _____
9. _____
10. _____
11. _____
12. _____
13. _____

34. _____
33. _____
32. _____
31. _____
30. _____
29. _____
28. _____
27. _____
26. _____
25. _____
24. _____
23. _____
22. _____
21. _____
20. _____
19. _____
18. _____
17. _____
16. _____
15. _____
14. _____

Muscular System—
Anterior View

Answers

Muscular System—Posterior View

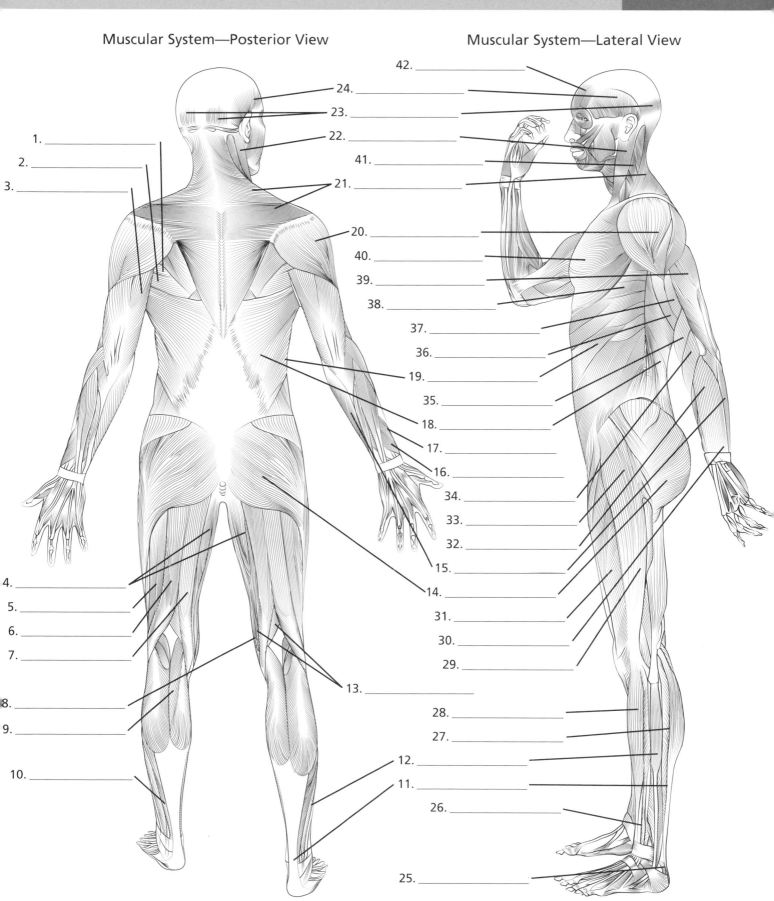

1. _____
2. _____
3. _____

4. _____
5. _____
6. _____
7. _____

8. _____
9. _____

10. _____

24. _____
23. _____
22. _____
21. _____

20. _____
19. _____
18. _____
17. _____
16. _____
15. _____
14. _____

13. _____

12. _____
11. _____

Muscular System—Lateral View

42. _____
41. _____
40. _____
39. _____
38. _____
37. _____
36. _____
35. _____
34. _____
33. _____
32. _____
31. _____
30. _____
29. _____

28. _____
27. _____
26. _____

25. _____

Answers

Muscles of the Thorax and Abdomen

Superficial and Deep Muscles
of the Thorax and Abdomen—
Anterior View

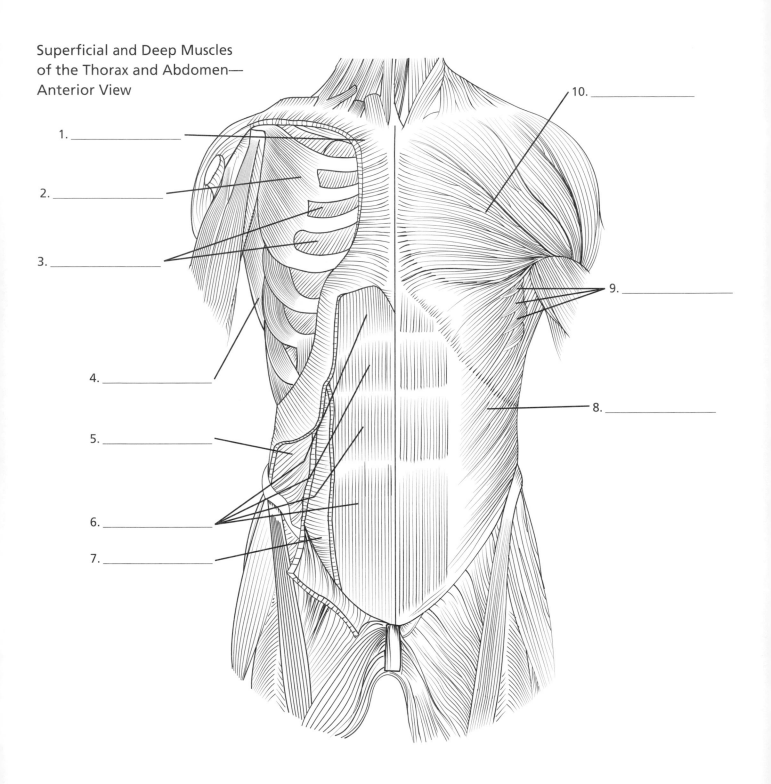

1. _____

2. _____

3. _____

4. _____

5. _____

6. _____

7. _____

8. _____

9. _____

10. _____

Answers

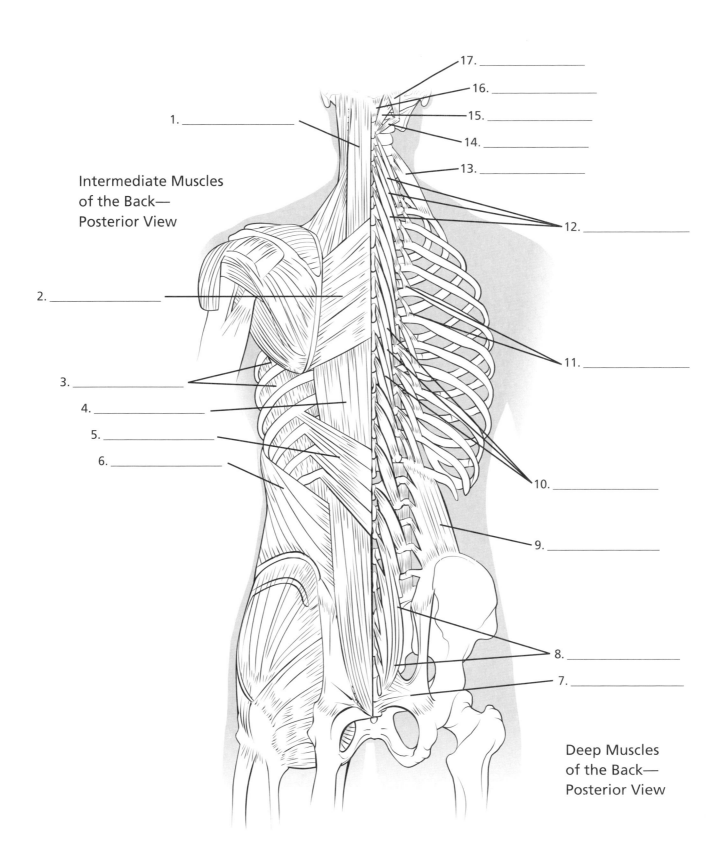

Intermediate Muscles
of the Back—
Posterior View

1. _____

2. _____

3. _____

4. _____

5. _____

6. _____

7. _____

8. _____

9. _____

10. _____

11. _____

12. _____

13. _____

14. _____

15. _____

16. _____

17. _____

Deep Muscles
of the Back—
Posterior View

Answers

1. Semispinalis capitis, 2. Rhomboid major, 3. External intercostals, 4. Erector spinae, 5. Serratus posterior inferior, 6. Internal oblique, 7. Sacrotuberous ligament, 8. Multifidus, 9. Quadratus lumborum, 10. Levatores costarum, 11. Semispinalis thoracis, 12. Semispinalis cervicis, 13. Scalenus posterior, 14. Obliquus capitis inferior, 15. Rectus capitis posterior major, 16. Rectus capitis posterior minor, 17. Obliquus capitis superior.

Muscles of the Upper Limb

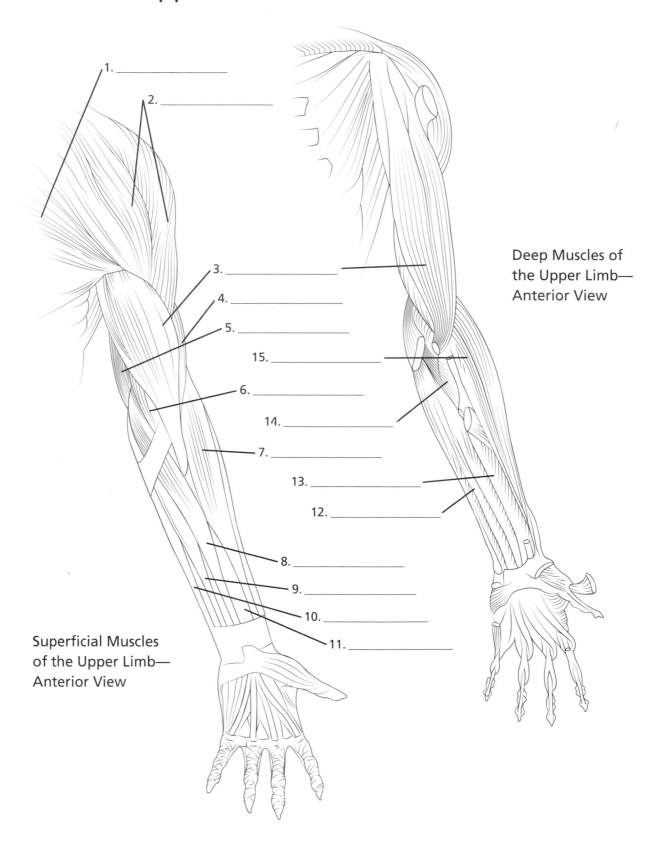

1. _____

2. _____

3. _____

4. _____

5. _____

15. _____

6. _____

14. _____

7. _____

13. _____

12. _____

8. _____

9. _____

10. _____

11. _____

Deep Muscles of
the Upper Limb—
Anterior View

Superficial Muscles
of the Upper Limb—
Anterior View

Answers

10. Tendon of flexor carpi ulnaris, 11. Flexor digitorum superficialis, 12. Flexor pollicis longus, 13. Flexor digitorum profundus, 14. Pronator teres, 15. Extensor carpi radialis longus

1. Pectoralis major, 2. Deltoid, 3. Biceps brachii, 4. Brachialis, 5. Triceps brachii, 6. Pronator teres, 7. Brachioradialis, 8. Tendon of flexor carpi radialis, 9. Tendon of palmaris longus,

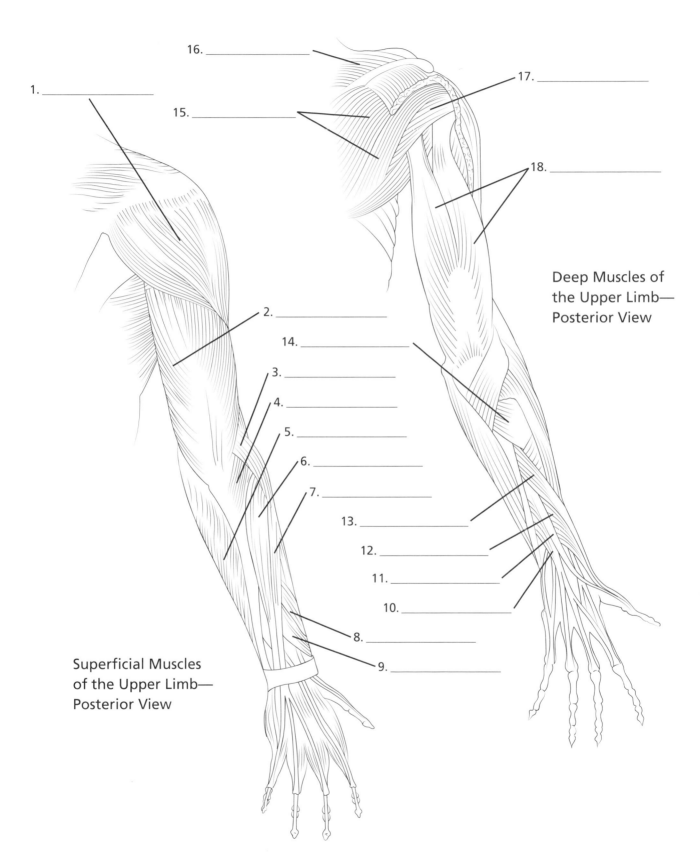

16. _____

1. _____

17. _____

15. _____

18. _____

Deep Muscles of
the Upper Limb—
Posterior View

2. _____

14. _____

3. _____

4. _____

5. _____

6. _____

7. _____

13. _____

12. _____

11. _____

10. _____

8. _____

9. _____

Superficial Muscles
of the Upper Limb—
Posterior View

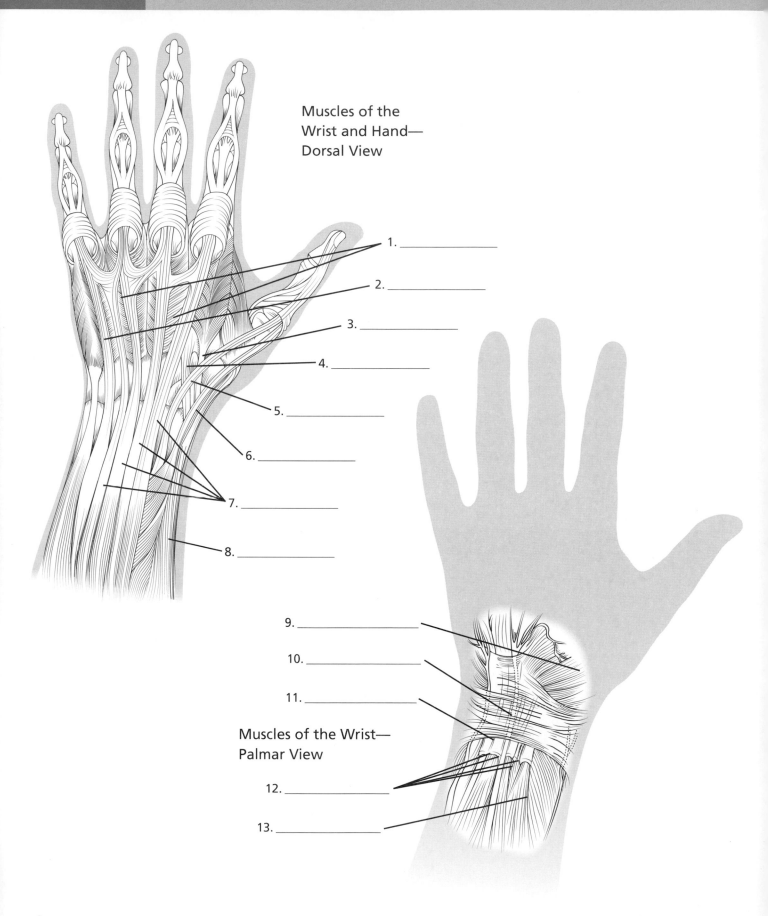

Muscles of the
Wrist and Hand—
Dorsal View

1. _____

2. _____

3. _____

4. _____

5. _____

6. _____

7. _____

8. _____

9. _____

10. _____

11. _____

Muscles of the Wrist—
Palmar View

12. _____

13. _____

Answers

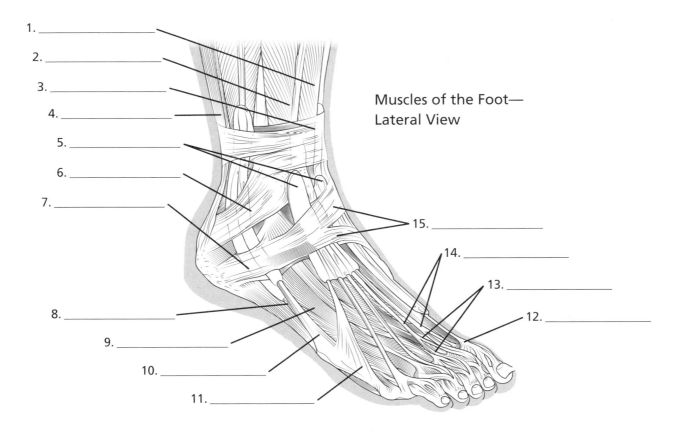

Muscles of the Foot—
Lateral View

1. _____
2. _____
3. _____
4. _____
5. _____
6. _____
7. _____
8. _____
9. _____
10. _____
11. _____

15. _____
14. _____
13. _____
12. _____

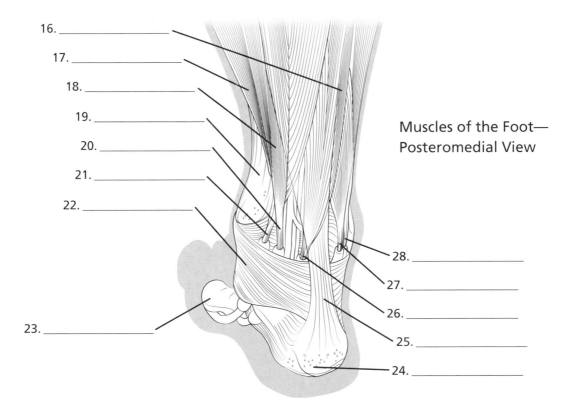

Muscles of the Foot—
Posteromedial View

16. _____
17. _____
18. _____
19. _____
20. _____
21. _____
22. _____
23. _____

28. _____
27. _____
26. _____
25. _____
24. _____

Answers

Muscles of the Lower Limb

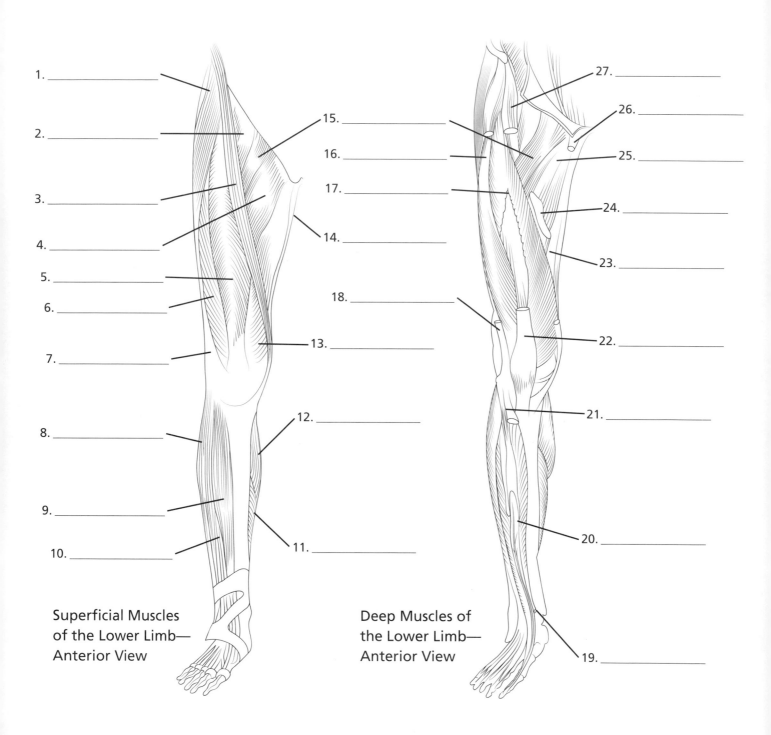

1. _____

2. _____

3. _____

4. _____

5. _____

6. _____

7. _____

8. _____

9. _____

10. _____

15. _____

16. _____

17. _____

14. _____

18. _____

13. _____

12. _____

11. _____

27. _____

26. _____

25. _____

24. _____

23. _____

22. _____

21. _____

20. _____

19. _____

Superficial Muscles of the Lower Limb— Anterior View

Deep Muscles of the Lower Limb— Anterior View

Answers

1. Tensor fasciae latae, 2. Iliopsoas, 3. Sartorius, 4. Adductor longus, 5. Rectus femoris, 6. Vastus lateralis, 7. Iliotibial tract, 8. Fibularis (peroneus) longus, 9. Tibialis anterior, 10. Extensor digitorum longus, 11. Soleus, 12. Gastrocnemius, 13. Pectineus, 14. Gracilis, 15. Vastus medialis, 16. Vastus lateralis, 17. Vastus intermedius, 18. Iliotibial tract (cut), 19. Tibialis anterior (cut), 20. Extensor hallucis longus, 21. Tibialis anterior (cut), 22. Rectus femoris (cut), 23. Adductor magnus, 24. Adductor longus (cut), 25. Adductor brevis, 26. Adductor longus (cut), 27. Sartorius (cut)

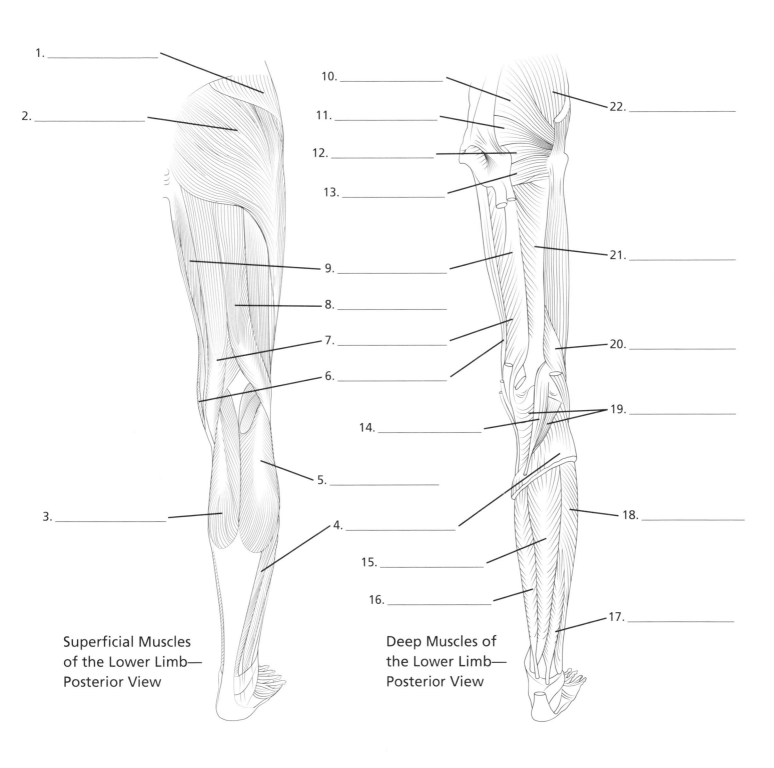

1. _____

2. _____

10. _____

11. _____

12. _____

13. _____

22. _____

9. _____

8. _____

7. _____

6. _____

14. _____

5. _____

4. _____

21. _____

20. _____

19. _____

18. _____

15. _____

16. _____

17. _____

3. _____

Superficial Muscles
of the Lower Limb—
Posterior View

Deep Muscles of
the Lower Limb—
Posterior View

Answers

1. Gluteus medius, 2. Gluteus maximus, 3. Medial head of gastrocnemius, 4. Soleus, 5. Lateral head of gastrocnemius, 6. Gracilis, 7. Semitendinosus, 8. Biceps femoris, 9. Adductor magnus, 10. Piriformis, 11. Superior gemellus, 12. Inferior gemellus, 13. Quadratus femoris, 14. Plantaris, 15. Tibialis posterior, 16. Flexor digitorum longus, 17. Flexor hallucis longus, 18. Fibularis (peroneus) longus, 19. Popliteus, 20. Short head of biceps femoris, 21. Adductor part of adductor magnus, 22. Gluteus minimus

Muscle Types

1. _____

2. _____

3. _____

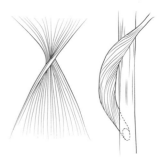

4. _____

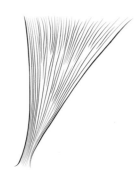

5. _____

6. _____

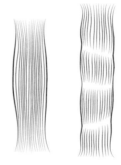

7. _____

8. _____

9. _____

10. _____

11. _____

12. _____

13. _____

14. _____

15. _____

16. _____

Answers

1. Unipennate, 2. Bipennate, 3. Multipennate, 4. Spiral, 5. Radial, 6. Quadrate, 7. Strap, 8. Cruciate, 9. Triangular, 10. Multicaudal, 11. Fusiform, 12. Digastric, 13. Circular (sphincteric), 14. Bicipital, 15. Tricipital, 16. Quadricipital

Articulations

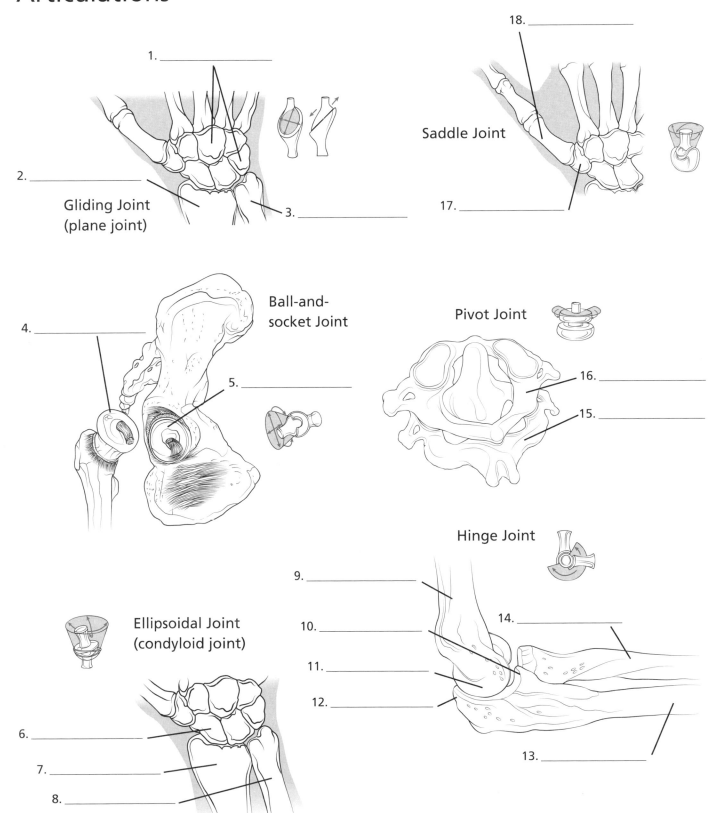

1. _____

2. _____

Gliding Joint
(plane joint)

3. _____

18. _____

Saddle Joint

17. _____

Ball-and-
socket Joint

4. _____

5. _____

Pivot Joint

16. _____

15. _____

Hinge Joint

9. _____

10. _____

11. _____

12. _____

14. _____

13. _____

Ellipsoidal Joint
(condyloid joint)

6. _____

7. _____

8. _____

Answers

1. Carpal bones, 2. Radius, 3. Ulna, 4. Head of femur (ball), 5. Acetabulum (socket), 6. Scaphoid bone, 7. Radius, 8. Ulna, 9. Humerus, 10. Coronoid process of ulna, 11. Trochlea (of humerus), 12. Olecranon, 13. Ulna, 14. Radius, 15. Axis, 16. Atlas, 17. Trapezium bone, 18. Metacarpal bone of thumb

Skeletal System

Skeletal System—
Anterior View

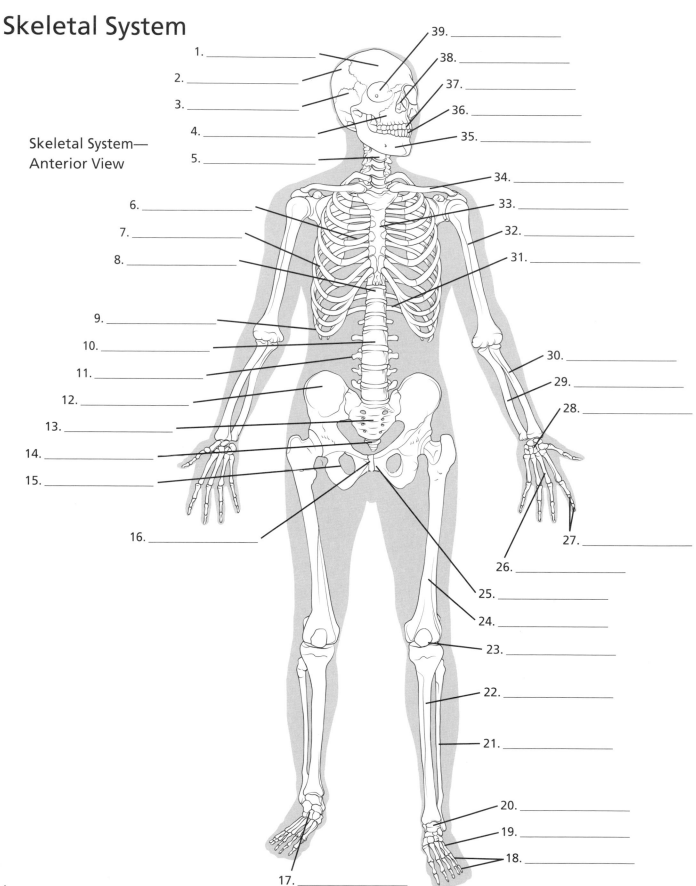

1. _____

2. _____

3. _____

4. _____

5. _____

6. _____

7. _____

8. _____

9. _____

10. _____

11. _____

12. _____

13. _____

14. _____

15. _____

16. _____

17. _____

39. _____

38. _____

37. _____

36. _____

35. _____

34. _____

33. _____

32. _____

31. _____

30. _____

29. _____

28. _____

27. _____

26. _____

25. _____

24. _____

23. _____

22. _____

21. _____

20. _____

19. _____

18. _____

Answers

Skeletal System—Posterior View

Skeletal System—Lateral View

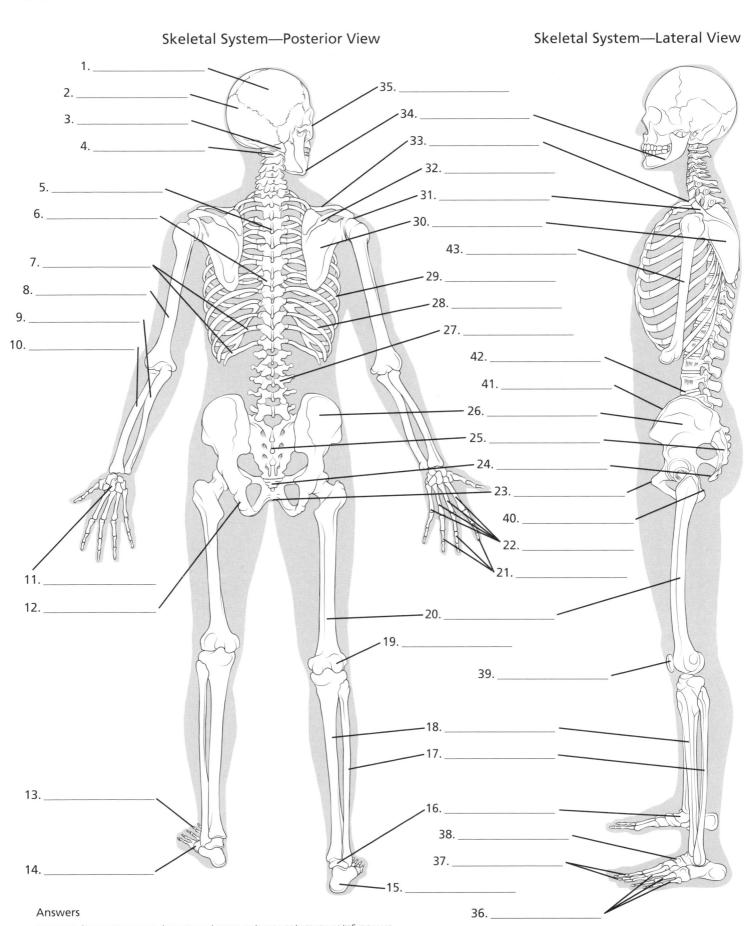

1. _____
2. _____
3. _____
4. _____
5. _____
6. _____
7. _____
8. _____
9. _____
10. _____
11. _____
12. _____
13. _____
14. _____

35. _____
34. _____
33. _____
32. _____
31. _____
30. _____
43. _____
29. _____
28. _____
27. _____
42. _____
41. _____
26. _____
25. _____
24. _____
23. _____
40. _____
22. _____
21. _____
20. _____
19. _____
39. _____
18. _____
17. _____
16. _____
38. _____
37. _____
36. _____
15. _____

Answers

1. Parietal bone, 2. Occipital bone, 3. Atlas (C1), 4. Axis (C2), 5. Spinous process of thoracic vertebra, 6. Thoracic vertebra, 7. Floating ribs (11 & 12), 8. Humerus, 9. Ulna, 10. Radius, 11. Carpal bones, 12. Ischial tuberosity, 13. Phalanges, 14. Metatarsal bones, 15. Calcaneus, 16. Talus, 17. Fibula, 18. Tibia, 19. Femoral condyle, 20. Femur, 21. Phalanges, 22. Metacarpal bones, 23. Pubis, 24. Coccyx, 25. Sacrum, 26. Ilium, 27. Lumbar vertebra, 28. False rib, 29. True rib, 30. Scapula, 31. Acromion, 32. Spine of the scapula, 33. Clavicle, 34. Mandible, 35. Zygomatic bone, 36. Metatarsal bones, 37. Phalanges, 38. Navicular, 39. Patella, 40. Ischium, 41. Iliac crest, 42. Intervertebral disk, 43. Humerus

Vertebral Column

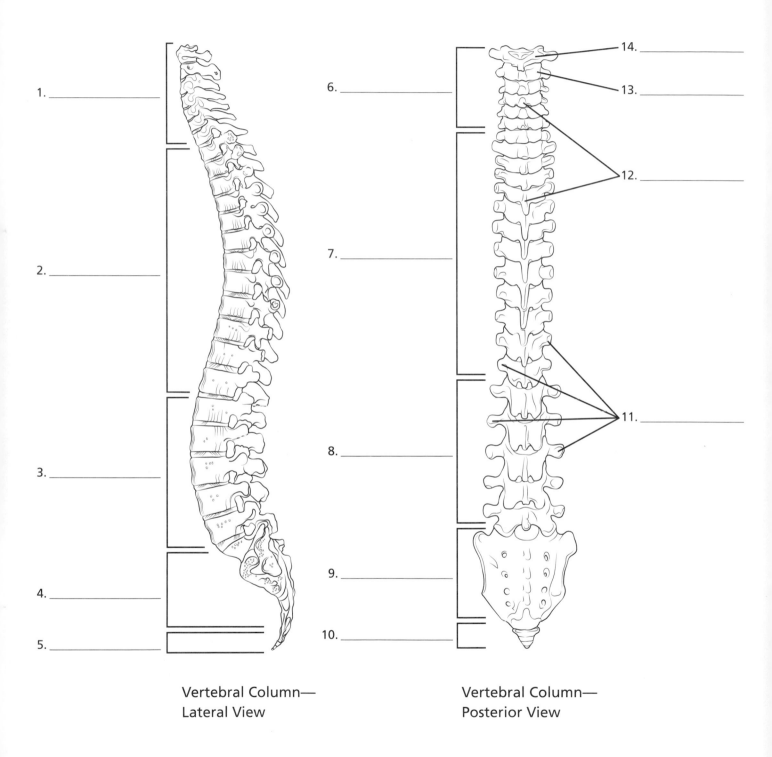

1. _____

2. _____

3. _____

4. _____

5. _____

6. _____

7. _____

8. _____

9. _____

10. _____

11. _____

12. _____

13. _____

14. _____

Vertebral Column—
Lateral View

Vertebral Column—
Posterior View

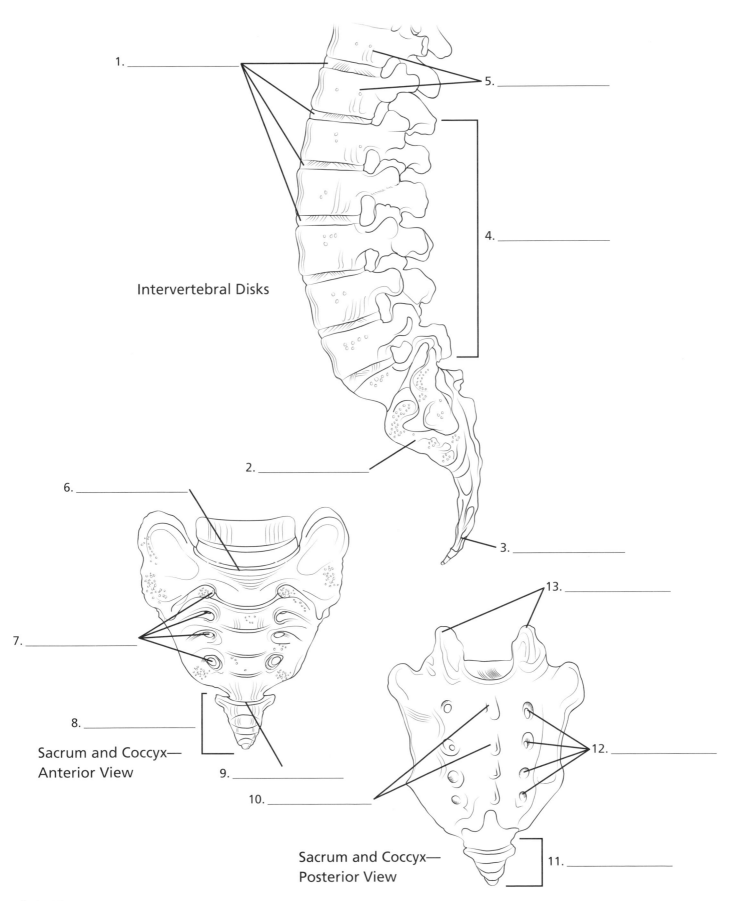

1. _____

5. _____

Intervertebral Disks

4. _____

2. _____

6. _____

3. _____

7. _____

13. _____

8. _____

12. _____

Sacrum and Coccyx—
Anterior View

9. _____

10. _____

11. _____

Sacrum and Coccyx—
Posterior View

Answers

Bones of the Upper Limb

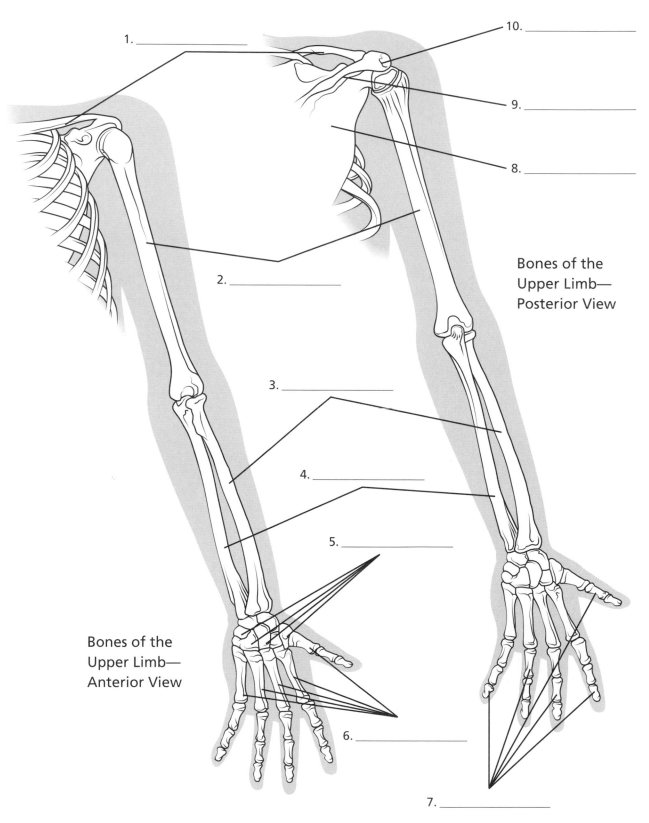

1. _____

10. _____

9. _____

8. _____

2. _____

3. _____

4. _____

5. _____

6. _____

7. _____

Bones of the
Upper Limb—
Posterior View

Bones of the
Upper Limb—
Anterior View

Answers

1. Clavicle, 2. Humerus, 3. Radius, 4. Ulna, 5. Carpal bones, 6. Metacarpal bones, 7. Phalanges, 8. Scapula, 9. Spine of scapula, 10. Acromion

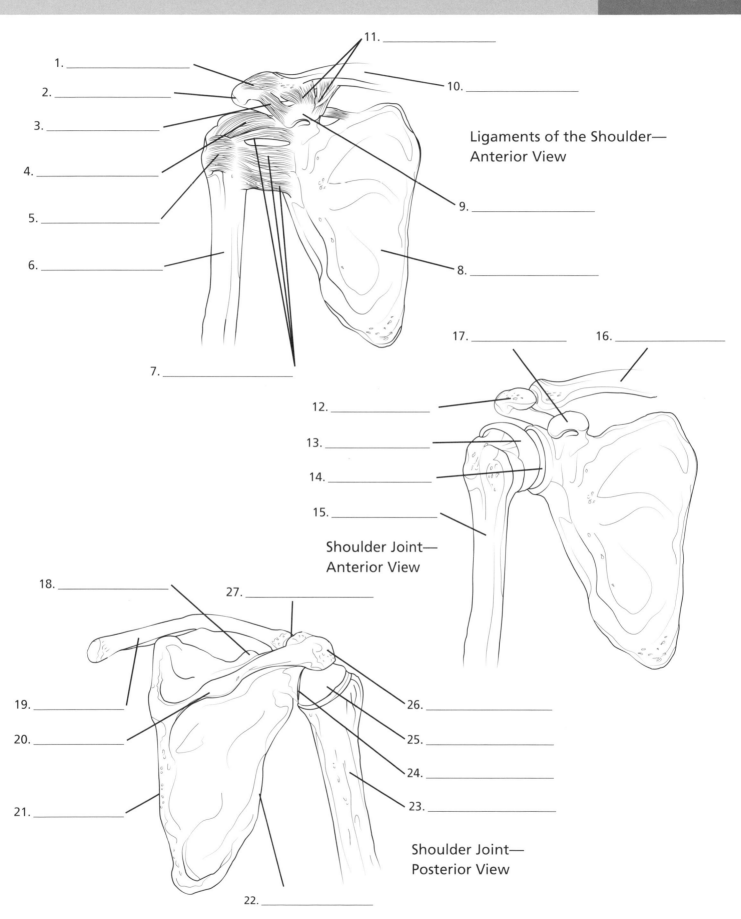

1. _____

2. _____

3. _____

4. _____

5. _____

6. _____

7. _____

11. _____

10. _____

9. _____

8. _____

Ligaments of the Shoulder—
Anterior View

17. _____

16. _____

12. _____

13. _____

14. _____

15. _____

Shoulder Joint—
Anterior View

18. _____

27. _____

19. _____

20. _____

21. _____

26. _____

25. _____

24. _____

23. _____

22. _____

Shoulder Joint—
Posterior View

Answers

1. Acromioclavicular ligament, 2. Acromion, 3. Coracoacromial ligament, 4. Coracohumeral ligament, 5. Transverse humeral ligament, 6. Shaft of humerus, 7. Glenohumeral ligaments, 8. Scapula, 9. Coracoid process, 10. Clavicle, 11. Coracoclavicular ligament, 12. Acromion, 13. Head of humerus, 14. Glenoid cavity, 15. Humerus, 16. Clavicle, 17. Coracoid, 18. Coracoid process, 19. Clavicle, 20. Spine of scapula, 21. Medial border of scapula, 22. Lateral border of scapula, 23. Humerus, 24. Glenoid fossa, 25. Head of humerus, 26. Acromion, 27. Acromioclavicular joint

Elbow Joint—Medial View

6. _____

7. _____

8. _____

9. _____

5. _____

4. _____

1. _____

2. _____

3. _____

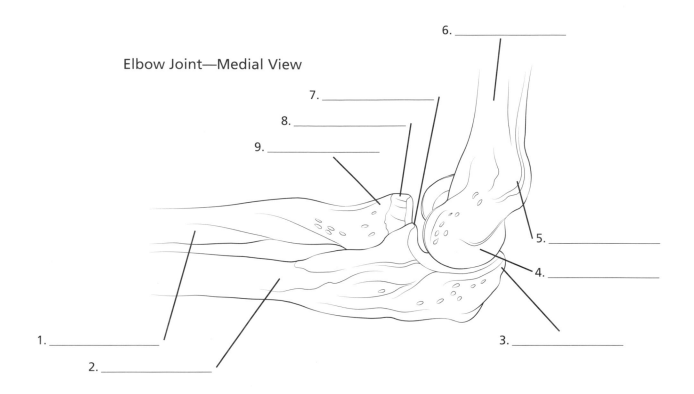

16. _____

Ligaments of the Elbow—Medial View

10. _____

11. _____

12. _____

13. _____

15. _____

14. _____

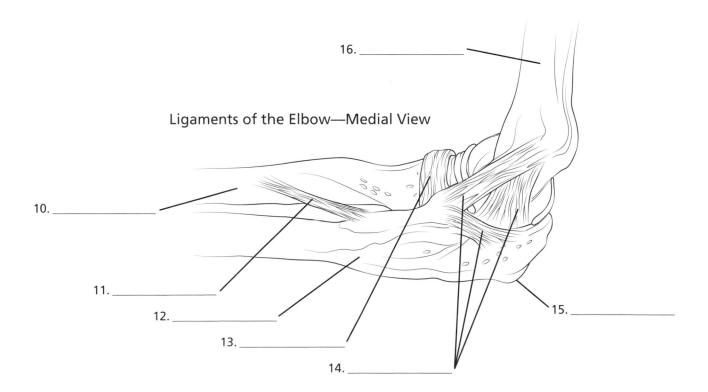

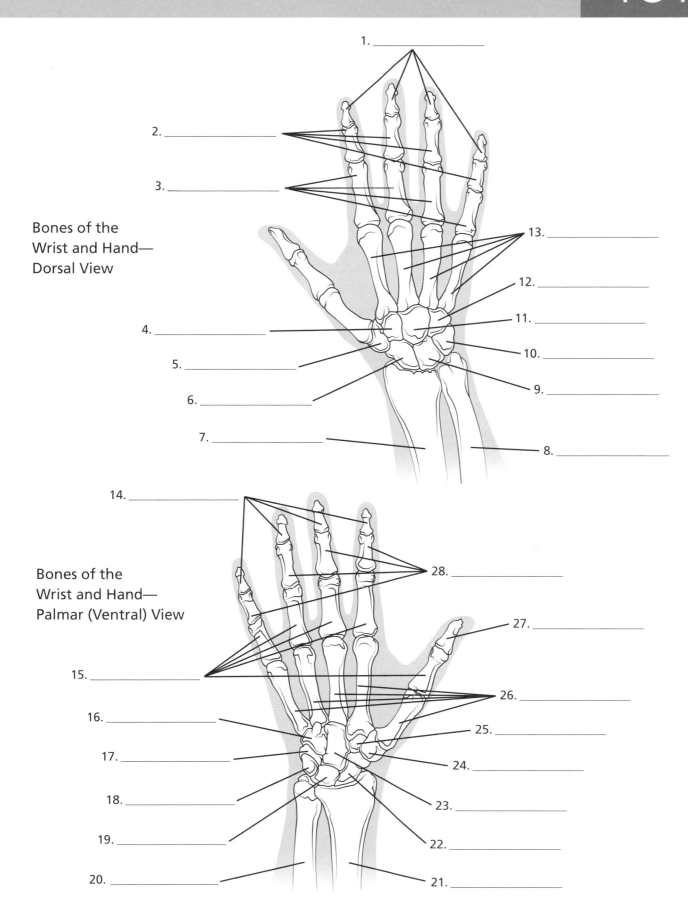

Bones of the
Wrist and Hand—
Dorsal View

1. _____
2. _____
3. _____
4. _____
5. _____
6. _____
7. _____
8. _____
9. _____
10. _____
11. _____
12. _____
13. _____

Bones of the
Wrist and Hand—
Palmar (Ventral) View

14. _____
15. _____
16. _____
17. _____
18. _____
19. _____
20. _____
21. _____
22. _____
23. _____
24. _____
25. _____
26. _____
27. _____
28. _____

Answers

1. Distal phalanges, 2. Middle phalanges, 3. Proximal phalanges, 4. Trapezoid, 5. Trapezium, 6. Scaphoid, 7. Radius, 8. Ulna, 9. Lunate, 10. Triquetrum, 11. Capitate, 12. Hamate, 13. Metacarpal bones, 14. Distal phalanges, 15. Proximal phalanges, 16. Hamate, 17. Triquetrum, 18. Pisiform, 19. Lunate, 20. Ulna, 21. Radius, 22. Scaphoid, 23. Capitate, 24. Trapezoid, 25. Trapezium, 26. Metacarpal bones, 27. Distal phalanx of thumb, 28. Middle phalanges

Bones of the Lower Limb

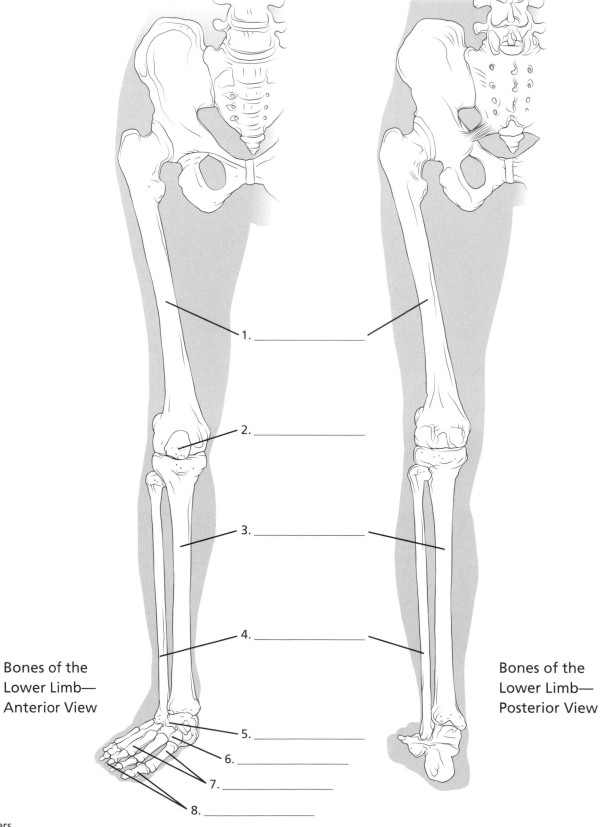

1. _____

2. _____

3. _____

4. _____

Bones of the
Lower Limb—
Anterior View

5. _____

6. _____

7. _____

8. _____

Bones of the
Lower Limb—
Posterior View

Answers

1. Femur, 2. Patella, 3. Tibia, 4. Fibula, 5. Talus, 6. Tarsal bones, 7. Metatarsal bones, 8. Phalanges

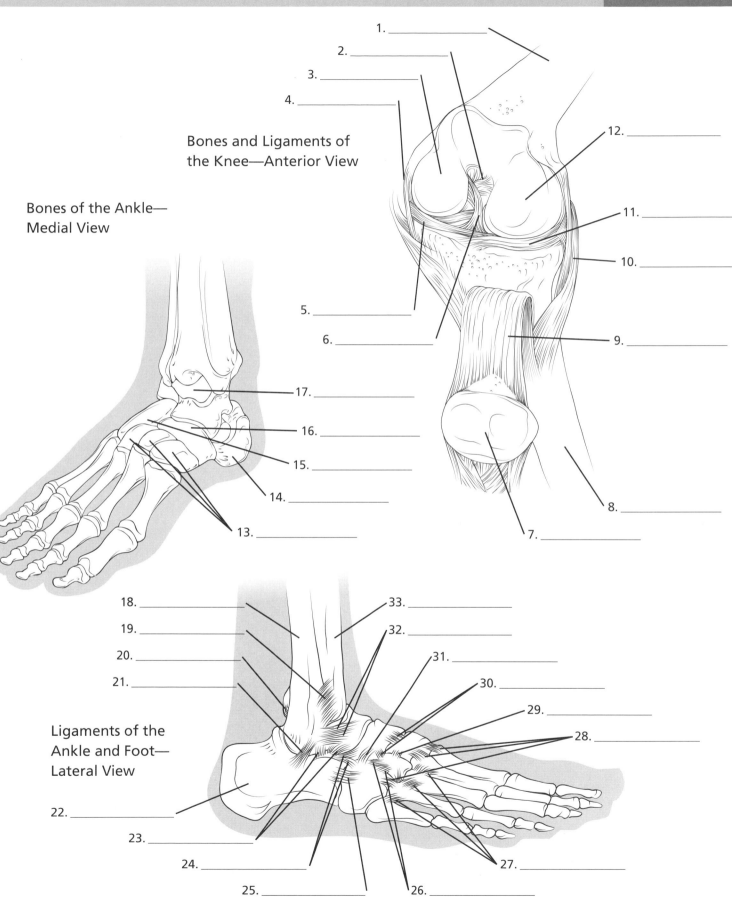

Bones and Ligaments of
the Knee—Anterior View

Bones of the Ankle—
Medial View

1. _____
2. _____
3. _____
4. _____
12. _____
11. _____
10. _____
5. _____
6. _____
9. _____
8. _____
7. _____
17. _____
16. _____
15. _____
14. _____
13. _____

Ligaments of the
Ankle and Foot—
Lateral View

18. _____
19. _____
20. _____
21. _____
22. _____
23. _____
24. _____
25. _____
26. _____
27. _____
28. _____
29. _____
30. _____
31. _____
32. _____
33. _____

Answers

1. Femur, 2. Posterior cruciate ligament, 3. Lateral condyle of femur, 4. Fibular (lateral) collateral ligament, 5. Lateral meniscus, 6. Anterior cruciate ligament, 7. Patella (reflected), 8. Tibia,
9. Patellar ligament, 10. Tibial (medial) collateral ligament, 11. Medial meniscus, 12. Cuneiform bones, 13. Cuneiform bones, 14. Calcaneus, 15. Cuboid, 16. Navicular, 17. Talus,
18. Fibula, 19. Anterior tibiofibular ligament, 20. Posterior tibiofibular ligament, 21. Calcaneofibular ligament, 22. Calcaneus, 23. Talocalcaneal ligaments, 24. Bifurcate ligament,
25. Dorsal calcaneocuboid ligament, 26. Dorsal cuneocuboid ligament, 27. Dorsal tarsometatarsal ligament, 28. Dorsal metatarsal ligaments, 29. Dorsal intercuneiform ligament,
30. Dorsal cuneonavicular ligaments, 31. Dorsal cuboideonavicular ligament, 32. Anterior talofibular ligament, 33. Tibia

Nerves of the Upper and Lower Limb

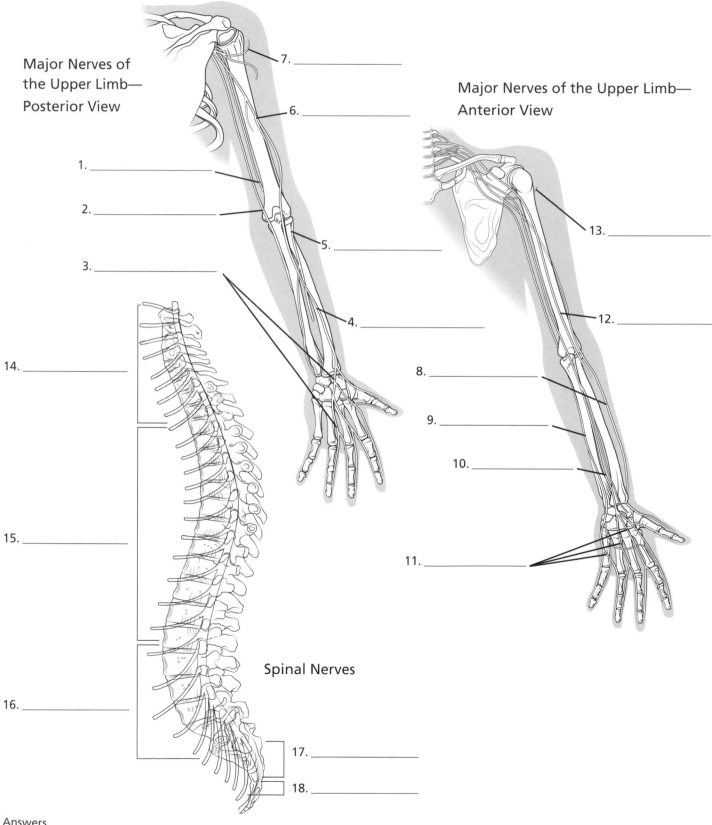

Major Nerves of
the Upper Limb—
Posterior View

7. _____

6. _____

1. _____

2. _____

5. _____

3. _____

4. _____

14. _____

15. _____

16. _____

Spinal Nerves

17. _____

18. _____

Major Nerves of the Upper Limb—
Anterior View

13. _____

12. _____

8. _____

9. _____

10. _____

11. _____

Major Nerves of the Lower Limb— Posterior View

Major Nerves of the Lower Limb— Anterior View

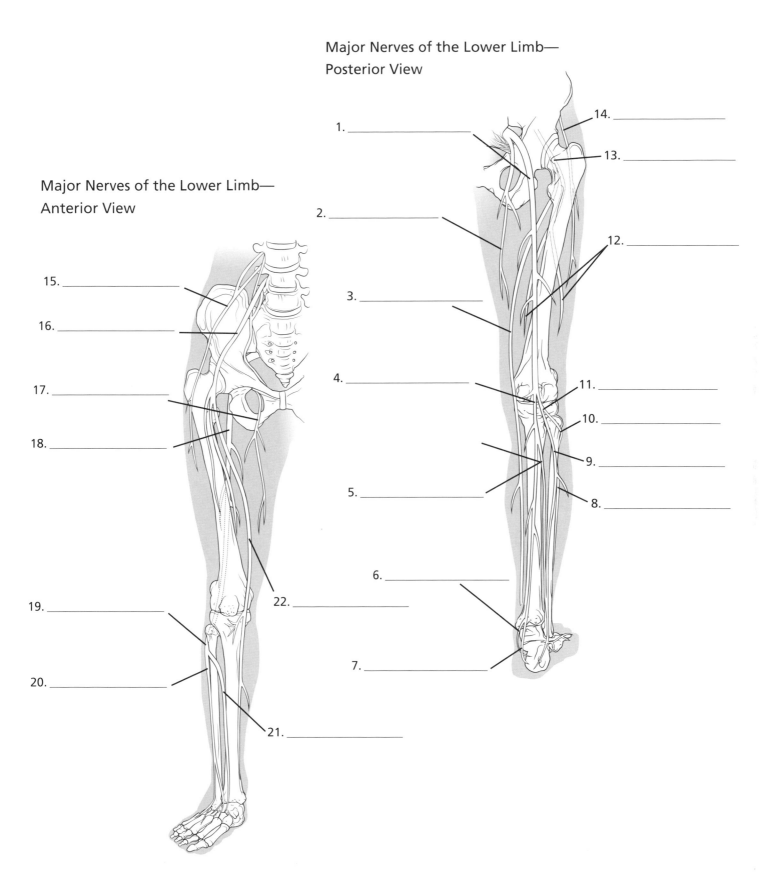

1. _____
2. _____
3. _____
4. _____
5. _____
6. _____
7. _____
8. _____
9. _____
10. _____
11. _____
12. _____
13. _____
14. _____

15. _____
16. _____
17. _____
18. _____
19. _____
20. _____
21. _____
22. _____

Answers

1. Sciatic nerve, 2. Posterior femoral cutaneous nerve, 3. Saphenous nerve, 4. Tibial nerve, 5. Medial sural cutaneous nerve, 6. Medial plantar nerve, 7. Lateral plantar nerve, 8. Lateral sural cutaneous nerve, 9. Deep fibular (peroneal) nerve, 10. Superficial fibular (peroneal) nerve, 11. Common fibular (peroneal) nerve, 12. Branches from femoral nerve, 13. Femoral nerve, 14. Lateral femoral cutaneous nerve, 15. Lateral femoral cutaneous nerve, 16. Femoral nerve, 17. Obturator nerve, 18. Sciatic nerve, 19. Common fibular (peroneal) nerve, 20. Superficial fibular (peroneal) nerve, 21. Deep fibular (peroneal) nerve, 22. Saphenous nerve

Reference

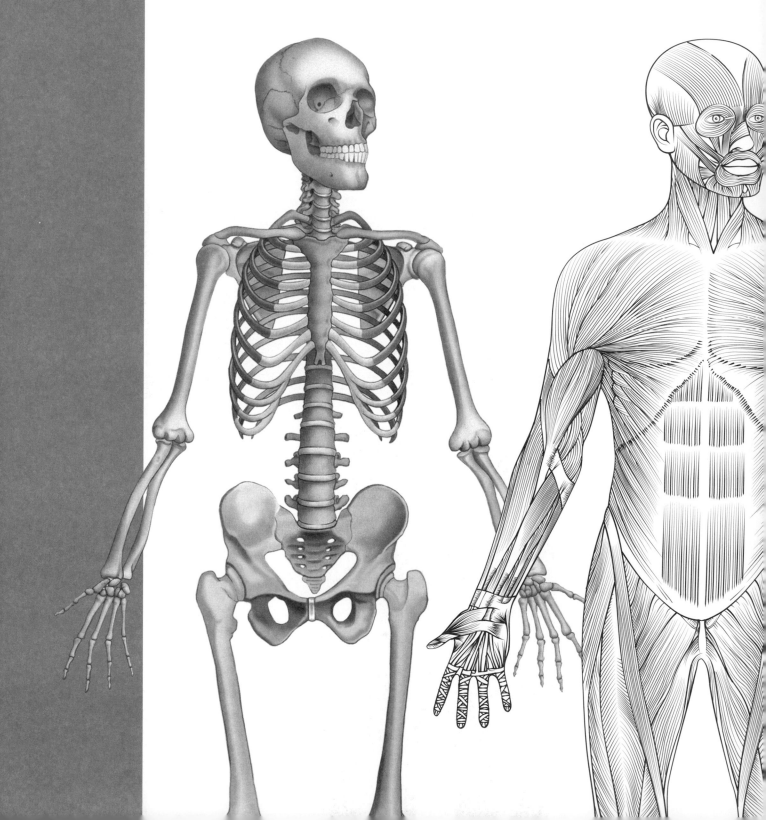

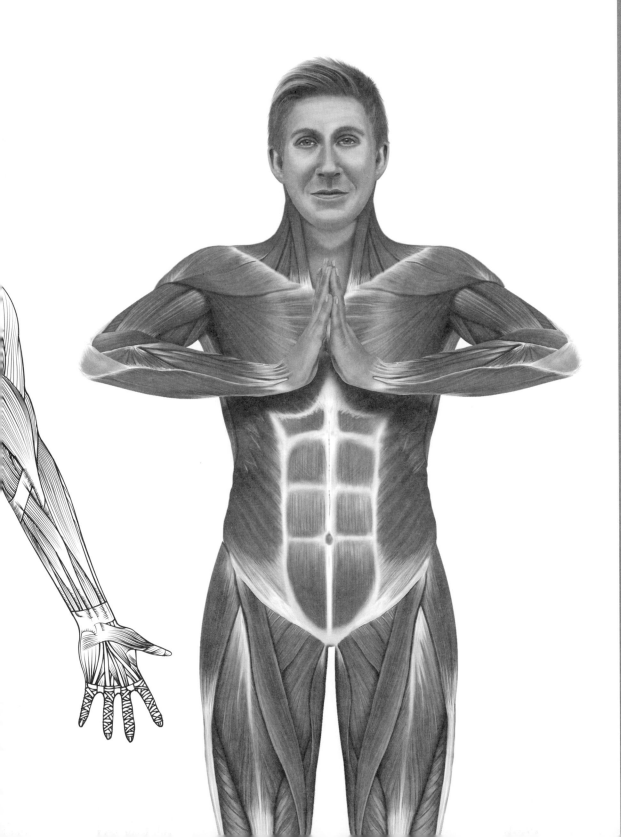

Glossary

Abduction Movement of a limb away from the midline of the body.

Accessory muscles *see* secondary muscles or movers.

Active muscles Muscles responsible for the main movement involved in an exercise or stretch. Also known as primary movers.

Adduction Movement of a limb toward the midline of the body.

Adductor muscles A group of muscles consisting of the adductor magnus, adductor longus, and adductor brevis. They lie on the inner side of the thigh.

Antagonist A muscle or group of muscles that bring about an opposite movement to the main action being considered.

Anterior Toward the front of the body.

Ballistic A type of stretching that uses the momentum of a moving body part to stretch a muscle into an extended range of movement that may exceed its static stretching range.

Buttocks (muscles) This group includes the gluteus maximus on the surface, and the gluteus medius and minimus muscles beneath.

Cervical vertebrae The seven vertebrae of the neck. Together they form a curve that is concave toward the back of the body.

Contractions The process of shortening a muscle. Produced by interaction between actin and myosin components of the myofibrils. They are initiated by an impulse from a motorneuron.

Core The trunk. Often used in reference to core stabilizing muscles (e.g., transverse abdominis, multifidus, obliques).

Core stability exercises Exercises for the muscles of the trunk that are key to maintaining good posture and balance, and protecting the internal organs.

Delayed onset muscle soreness (DOMS) A syndrome of muscular pain that occurs some hours to days after an unaccustomed or strenuous activity.

Dorsiflexion Used in the context of the ankle, it refers to the movement of the toes and top of the foot toward the head by bending at the ankle.

Dynamic stretching A type of stretching that uses the momentum of a moving body part to stretch a muscle into an extended range of movement that does not exceed its static stretching range.

Extension The act of straightening a limb at a joint.

Extensor muscles A group of muscles that perform an extension movement.

Flexion The act of bending a limb at a joint.

Flexor muscles A group of muscles that perform a flexion movement.

Hamstrings Muscles of the back of the thigh. The hamstring muscle group includes the semitendinosus, semimembranosus, and biceps femoris muscles.

Hyperextension Extension of a joint beyond the normal range of movement.

Internal impingement Used in the context of a nerve. Refers to a situation where a nerve is caught or pinched over a bony feature or hard ligament, or trapped by soft tissue adhesions.

Internal rotation Rotating a limb toward the midline. Also called medial rotation.

Kyphosis A curvature of the vertebral column that is concave to the front of the body. It is normal in the thoracic spine.

Lordosis A curvature of the vertebral column that is concave to the back of the body. It is normal in the lumbar and cervical spine.

Lumbar vertebrae The five vertebrae between the bottom of the rib cage and the pelvis. They comprise the lower back.

Medial rotation *see* internal rotation.

Mover muscles Muscles that work intermittently to produce a specific movement, such as latissimus dorsi and biceps brachii.

Neural glide technique A form of manipulative therapy and/or stretching that helps mobilize a nerve that may have been caught by adjacent soft or hard tissue.

Neutral position The natural position of the body or a specific body part.

Neutral spine The position of the vertebral column or spine, where there is the least stress on joints, ligaments, and disks. In the lumbar spine, this is a position of slight lordosis.

Plantar-flexor A muscle that produces a downward movement of the foot by movement at the ankle joint (e.g., the gastrocnemius).

PNF (proprio-neuromuscular-facilitation) technique A stretching technique that involves a shortening contraction of an opposing muscle to place the target muscle on stretch, followed by an isometric (i.e., same length) contraction of the target muscle. The name is derived from the effects on the myotatic stretch reflex pathways between the muscles and spinal cord.

Posterior core Core stabilizers of the back (e.g., multifidus).

Postural muscles Muscles that work continuously throughout the day to maintain posture, such as multifidus in the back.

Primary mover The main muscle or muscles that produce a movement.

Pronate Rotating the hand so that the palm faces downwards.

Quadriceps The quadriceps femoris muscle group on the front of the thigh includes the vastus lateralis, vastus intermedius, vastus medialis, and rectus femoris muscles.

Retraction Backward movement of the scapula toward the vertebral column.

Rotator cuff muscles A group of muscles (supraspinatus, subscapularis, infraspinatus, and teres minor) that arise from the scapula and insert onto the humerus to provide dynamic stability for the shoulder joint.

Scapula The bone commonly known as the shoulder blade.

Spotter Someone who assists the stretcher to get into the correct starting position, and assists if the stretcher struggles to maintain balance.

Static Hold still. Used in the context of a stretch, it means a stretch in which limb parts do not move significantly.

Supinate Rotating the hand so that the palm faces upwards.

Sustained Held for a prolonged period (e.g., 30 seconds to several minutes) in the context of a stretch.

Thoracic vertebrae The 12 vertebrae of the chest, comprising the upper and middle back—each is attached to a rib.

Index